Intelligent Systems Reference Library

Volume 64

For further volumes:
http://www.springer.com/series/8578

About this Series

The aim of this series is to publish a Reference Library, including novel advances and developments in all aspects of Intelligent Systems in an easily accessible and well structured form. The series includes reference works, handbooks, compendia, textbooks, well-structured monographs, dictionaries, and encyclopedias. It contains well integrated knowledge and current information in the field of Intelligent Systems. The series covers the theory, applications, and design methods of Intelligent Systems. Virtually all disciplines such as engineering, computer science, avionics, business, e-commerce, environment, healthcare, physics and life science are included.

Efthimios Alepis · Maria Virvou

Object-Oriented User Interfaces for Personalized Mobile Learning

Efthimios Alepis
Maria Virvou
Department of Informatics
University of Piraeus
Piraeus
Greece

ISSN 1868-4394 ISSN 1868-4408 (electronic)
ISBN 978-3-662-50654-7 ISBN 978-3-642-53851-3 (eBook)
DOI 10.1007/978-3-642-53851-3
Springer Heidelberg New York Dordrecht London

Softcover reprint of the hardcover 1st edition 2014

Printed on acid-free paper

Springer is part of Springer Science+Business Media (www.springer.com)

Foreword

Efthimios Alepis and Maria Virvou have investigated two recent related areas that attracted the attention of the scientific community, namely mobile learning and interfaces. The motivation arose from the fact that, when integrating these technologies, we obtain personalized educational software that meets the prerequisites of modern mobile learning software that has become very popular worldwide in recent years. These two technologies have made significant advances recently and have become hot disciplines with increasing research projects around the world in both academia and industry.

Demand for mobile learning is growing at a remarkable rate; however, there seems to be a shortfall in software development to meet this fast-growing demand and associated challenges.

This book is a significant addition to this field and an excellent effort to address these challenges and trends. The authors employ an interesting approach that utilizes the Object-Oriented (OO) method in order to find answers for these issues and difficulties. They chose to follow the object-oriented scheme so as to embrace the basic concepts and traits in order to offer a very flexible, vigorous, and extendable structure for the devised framework.

Specifically, in the book, the authors develop a broad paradigm built using the OO approach. I found that each chapter concentrates on the structure of a particular section of the paradigm; however, it puts all of these together in a nice way.

I believe that the authors have done a good job at addressing the tackled issues. I consider the book a good addition to the areas of mobile learning and user interfaces. It definitely will help software developers to build better state-of-the-art personalized software aiming at mobile education, while maintaining a high level of adaptivity and user-friendliness within individualized-mobile interfaces.

New Jersey, USA Prof. Mohammad S. Obaidat

Preface

This book covers two very important and quite recent scientific fields, namely that of mobile learning and the other, advanced user interfaces. These two scientific fields' successful combination can result in personalized educational software that meets the requirements of state-of-the-art mobile learning software. Both mobile learning and user-personalized interfaces have grown over the last decade from minor research fields to a large set of significant projects in universities, schools, workplaces, museums, and cities around the world. According to a report in 2013, "the market for Mobile Learning products and services has been growing at a five-year compounded annual growth rate of more than 25 %." Benefits by using and/or incorporating these technologies in software engineering include social, economic, and educational gains. However, the swift growth of new software technologies and their corresponding services keeps in pace with new challenges in these scientific fields. As a result, new approaches try to resolve the resulting problems and at the same time give more potential and robustness to the next generation of software applications.

In this book, the authors try to provide a framework that is capable of incorporating the aforementioned software technologies, exploiting a wide range of their current advances and additionally investigates ways to go even further by providing potential solutions to future challenges. Our proposed approach uses the well-known Object-Oriented method in order to address these challenges. By using the OO approach, we adopt its fundamental concepts and features for the purposes of providing a highly adjustable, dynamic, and extendable architecture for our proposed framework. Throughout this book, a general model is constructed using Object-Oriented Architecture. Each chapter focuses on the construction of a specific part of this model, while in the conclusion these parts are unified. We believe that this book will help software engineers build more sophisticated personalized software that targets in mobile education, while at the same time retaining a high level of adaptivity and user-friendliness within human-mobile interaction.

Contents

Chapter 1
Introduction

Abstract In the first chapter of this book the authors present a short introduction on the scientific topics that are covered. To this end, this chapter includes introductory sections for the scientific fields of mobile education, multimodal mobile interfaces and the Object Oriented paradigm which, to a large extent, is followed throughout this book. Both mobile education and mobile multimodal user interfaces are quite recent and are rapidly growing fields in the broader areas of information technology. Their use is expected to grow at a serious rate in the foreseeable future, not only by technical, but also by software means of evolution. The introductory chapter also reveals the authors' motives for writing a book in the aforementioned domains while the rest of the book covers Object Oriented mobile projects, coupled by software evaluations and empirical studies.

1.1 Mobile Education

Over the last decade, both educators and educational institutions have recognized the importance of distance learning educational software. Some important assets of distance learning applications include platform, hardware and facilities independence and also the actual facility offered to students as distant learners of learning something at any time and at any place, away from the settings of a real classroom. In many situations this means that learning may take place at home or some other site, supervised remotely, synchronously or asynchronously, by human instructors as educators. In these cases the interaction is achieved between humans and computers through computer assisted learning or e-learning.

However, there are many cases where it would be extremely useful to have such facilities in portable handheld devices, rather than desktop or laptop computers so that users could use the educational software on a device that they can carry anywhere they go. Handheld devices may render the software usable on every occasion, even when people are waiting in a queue or even moving rather than when they are sitting on a chair inside their office. However, among handheld

E. Alepis and M. Virvou, *Object-Oriented User Interfaces for Personalized Mobile Learning*, Intelligent Systems Reference Library 64,
DOI: 10.1007/978-3-642-53851-3_1,

devices, which include palm or pocket PCs and mobile phones, mobile phones provide the additional very important asset of computer-device independence for users. Unlike mobile phones, palm-top PCs have to be bought by a person for the special purposes of computer use. On the other hand, mobile phones are very widespread devices, which are primarily used for human–human communication purposes. However, a large number of mobile devices can also be used as computers. Thus, prospective users of handheld devices are not required to spend money for extra computer equipment since they can use their mobile phone, which they would buy and carry with them anyway. In this sense, using the mobile phone as a handheld computer is a very cost-effective solution that provides many assets. Two of the most important assets are users' device independence as well as independence with respect to time and place in comparison with web-based or desktop-based education using standard PCs. Indeed, there are situations where both students and instructors could benefit by using their spare time constructively to complete homework exercises and electronic lesson authoring respectively in situations where no computer may be available. Such situations may occur in taxis, public buses and coaches while commuting, in long queues while waiting or more generally in situations where unexpected spare time comes up. In the fast pace of modern life where time is precious such situations can be very frequent. Mobile technology in general can provide services to most computer-based applications including educational applications. Mobile features can be of great assistance to educational procedures since they offer mobility to students and/or teachers.

In view of these compelling needs, the research work described in this book has dealt partially with the problem of enriching existing educational software technology with mobile aspects. In particular, in this chapter we focus on the incorporation of mobile educating and authoring tools into the field of Intelligent Tutoring Systems (ITSs). The work that is described in this chapter also resulted in the development of an authoring tool prototype that can generate ITSs of multiple domains. This authoring tool is called Mobile Author (Virvou and Alepis 2005). Mobile Author allows instructors to create and administer data-bases concerning characteristics of students, of the domain to be taught and of tests and homework. The creation and administration of these data-bases can be carried out through a user-friendly interface from any computer or mobile phone. In this way the creation of mobile ITSs is facilitated enormously and a high degree of reusability is ensured.

Authoring tools in general are meant to be used by human instructors (prospective e-learning authors) to build intelligent computerized tutors in a wide range of domains, including customer service, mathematics, equipment maintenance, and public policy; these tutors have been targeted toward a wide range of students, from grade school children to corporate trainees (Murray 1999). More specifically, authoring tools that specialize in ITSs aim at providing environments for cost-effective development of tutoring systems that can be intelligent and adaptive to individual students. The main goal of ITSs, as compared to other educational technologies, is to provide highly individualized guidance to students.

It is simple logic whose response, individualized to a particular student, must be based on some information about that student; in ITS technology this realization led to learner modeling, which became a core or even defining issue for the field (Cumming and McDougall 2000).

Similarly, in the resulting tutoring applications, students can answer test questions and can read parts of the theory from any computer or mobile phone. The underlying reasoning of the tutoring systems is based on the student modeling component of the resulting educational applications. The student modeling component monitors the students' actions while they use the educational system and tries to diagnose possible problems, recognize goals, record permanent habits and errors that are made repeatedly. The inferences made by the system concerning the students' characteristics are recorded in their student model that is used by the system to offer advice adapted to the needs of individual students. Moreover, the students' characteristics can be accessed by human instructors who may wish to see their students' progress and educational skills.

Despite the fact that computer-based mobile technology is quite recent, it is growing very rapidly and there have already been quite a few research attempts that aim at the incorporation of mobile features in education. As an example, Ketamo (2003) has developed an adaptive geometry game for handheld devices. Another approach is described in the system called KleOS (Vavoula and Sharples 2002). This system allows users to organize and manage their learning experiences and resources as a visual timeline in both desktop computers and mobile devices. An interesting application of mobile services in computer assisted education is the proposal of Wang et al. (2003). This proposal consists of three kinds of information awareness mechanisms using mobile devices to assist students in promoting their learning performances. Aiming at providing a more generalized framework, Leung and Chan (2003) introduce a framework of mobile learning that consists of mobile learning applications, mobile user infrastructure, mobile protocols and mobile network infrastructure. A common issue in all of the above systems is the fact that these systems aim primarily at assisting students in their learning process. However, there is also one important part of the educational process which refers to that part that teachers are involved in the educational processes and deals with creating and managing their courses and/or related educational material. Looking at this view, of computer assisted education, there are noticeable fewer attempts so far. For example (Chan et al. 2003) have made an attempt to address this issue. Their system aims at assisting teachers to create and manage their computer-based lessons.

In any case, it remains to be investigated quite thoroughly what the extent of the appreciation of these mobile features is within the educational community. Mobile features show a great potential to provide time, place and computer-equipment independence. On the other hand, mobile phones have more restricted interaction channels with users due to the limited space of memory capacity of the devices. This may not be a problem for simple uses of mobile phones but it may be a problem for the mobile use of more sophisticated computer-based applications such as educational applications. In view of these, in this book we also focus on the evaluation of the overall usefulness and usability of mobile educational applications. In Chap. 8 we will give a brief description of a mobile system and we will focus mainly on the usability aspects of its mobile features and we will also report on the primary results of an evaluation study that aimed at evaluating the usefulness of the mobile facilities and their usability.

1.2 The Object Oriented Paradigm

Software engineers' main pursuit is to create efficient, stable and effective software. The most well-known modern approach in designing software is through the Object Oriented Design. In this approach a software system is seen as a set of interacting objects. The purpose of their interaction is to solve specific problems in software engineering. Each object may be simple or composite and many objects together can built even the most complex software models.

OO Design is primarily a method in which programmers can realize what they create in a more understandable and physical way. This is because their world, as humans, is preferably also understood as entities or objects that interact with each other. A man (object) reads (does something) a book (another object). A second highly important reason for using the OO design is the advanced complexity of modern software. It is quite common in most complete software applications to observe millions of programming code comprising them and also numerous modules and services interacting with each other. In order to build, maintain and improve something with such a high complexity, the OO design does something that can be compared with the famous "divide-and-conquer" algorithmic approach. This approach works by recursively breaking down a complicated problem into more sub-problems of less complexity and of the same type, until they become simple enough to be solved. As a next step in this approach, the solutions to the sub-problems are finally combined to provide a solution for the original problem. As an example, it would be very difficult for someone to understand how a big organization works, by just watching the organization from the outside. However, if someone could observe how the people who work inside this organization do their work, this would give a clearer view of the organization in general.

In the same way, a complex software system cannot be de-synthesized by looking its GUI (graphical user interface) because we cannot presume its underlying mechanisms. A look in the system's code, which is usually thousands or millions of lines of code, cannot tell much more, even for experienced programmers. The OO architectural design comes to provide a symbolic way to illustrate most, if not all, of what is "inside" this fore mentioned system, in a way that today's humans find easy to comprehend.

To this end, the prevalence of modern programming languages such as Java, C++, C# and Object Pascal, make it very obvious that object-oriented design has become the approach of choice for most development projects in the world.

However, object oriented software developers should have a basic understanding in object-orientation in order to incorporate it in their future developments. Such understanding includes basic OO concepts such as inheritance, polymorphism, and abstraction.

Furthermore, OO software developers should have studied one of the most prevailing industrial standards in this domain, namely the Unified Modeling Language (UML). UML diagrams are most helpful even before proceeding to the OO design, in a phase known as Object-Oriented Analysis. With UML, diagrams can be created in order to illustrate the static structure, dynamic behavior, and run-time deployment of collaborating objects, defined in the analysis phase. The transition from these diagrams to an OO language (programming code) is then a straight forward process, easily carried through by most software engineers.

Throughout this book, outlines of some basic and most used UML diagrams are illustrated. Viewing a system from its internal scope, as software objects interacting with each other, class diagrams are highly recommended for this illustration and are thus used in most of this book's chapters. Viewing a system from the users' perspective, it is quite representative to illustrate the human-system interaction using "Use-Case" diagrams.

Figure 1.1 illustrates a class. A class consists of three parts, namely the class's name, the class's attributes and the class's operations. The concept of the class is inextricably connected to the concept of the object. Objects are instances of classes, while classes can be viewed as general templates. Figure 1.2 illustrates a system's class diagram in its primal general form where a system is comprised of subsystems and components. Child classes inherit from their basic parent class and we may also observe basic associations between classes and also other kinds of associations, such as aggregations and compositions. Finally, Fig. 1.3 illustrates an outline of a use case diagram, where there are three external actors interacting with the system and each one of them has its use cases.

1.3 Mobile Multimodal User Interfaces

As it is stated by many scientists in the related scientific field, next generation human–computer interaction (HCI) designs need to include the ability to recognize users' affective states in order to become more human-like, more adaptive, more effective and more efficient (Pantic and Rothkrantz 2003). In many situations human–computer interaction may be improved by achieving multimodal emotional interaction in real time (Jascana et al. 2008; Bernhardt and Robinson 2008). Affective computing has recently become a very important field of research as it focuses on recognizing and reproducing human feelings within human computer interaction. The skills of emotional intelligence have been argued to be a better predictor than IQ for measuring aspects of success in life (Goleman 1995). For this reason, human feelings are considered very important but only recently they have started being taken into account in human–computer interaction. Understanding

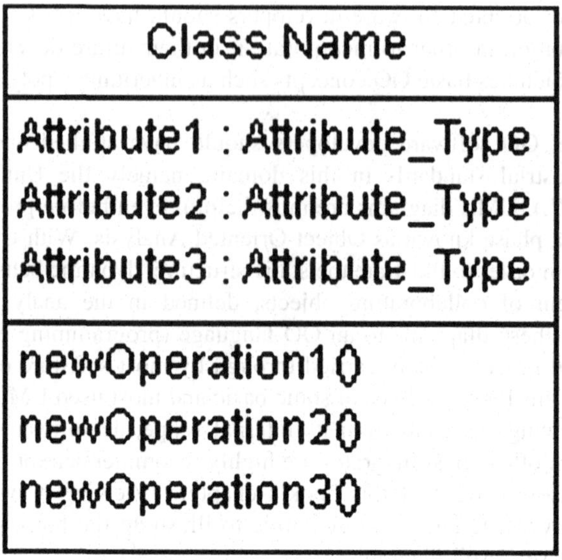

Class Name
Attribute1 : Attribute_Type Attribute2 : Attribute_Type Attribute3 : Attribute_Type
newOperation1() newOperation2() newOperation3()

Fig. 1.1 Representation of a class in UML

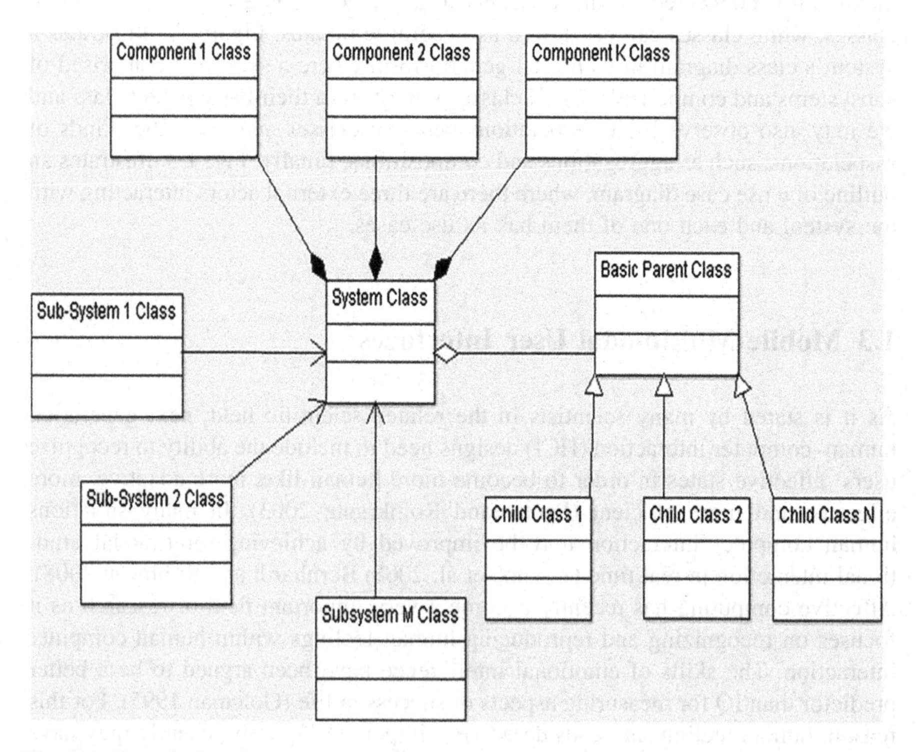

Fig. 1.2 General class diagram

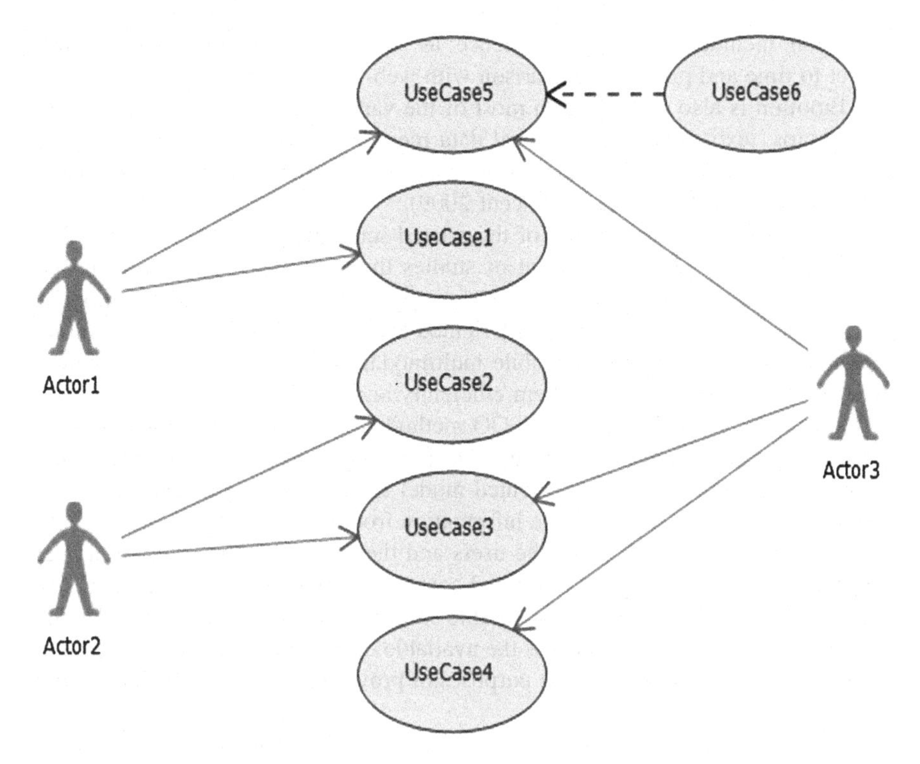

Fig. 1.3 General use case diagram

how human emotional communication exploits information from several different channels as modalities of interaction is quite demanding and requires the definition and constitution of a cross-modal and cross-cultural database comprising of verbal and non-verbal data (Esposito 2009). Thus, many scientists state that the area of affective computing is not yet well understood and explored and needs a large amount of research efforts to reach maturity.

In the last decade, there is also a growing interest in mobile user interfaces and mobile networks. It is estimated that over one half of the world's population has been using mobile phones since 2009 and many have become attached to and even dependent on these small electronic communication devices (Vincent 2009). As a result, a great number of services are offered to the users of mobile phones including education. In the fast pace of modern life, students and instructors would appreciate using constructively some spare time that would otherwise have been wasted (Virvou and Alepis 2005). Students and instructors may have to work on lessons, even when away from offices, classrooms and labs where computers are usually located. Correspondingly, in Virvou and Alepis (2004), Virvou and Alepis (2005) we have investigated how acceptable and useful the mobile features of an e-learning system have been to students that have used the system in comparison to the use of the system through a desktop computer. Advantages of mobile interaction include device independence as well as more independence with respect to time and place in comparison with web-based education using standard PCs. Emotion is also embedded in most of the various uses of the mobile phone, from photos, texts, and other personal data recorded on the mobile phone to the touch and feel of the device, and the association a particular mobile might have with a special event or person (Vincent 2009).

After a thorough investigation of the related scientific literature we found that there is a relatively limited amount of studies that aim at the data structures of multimodal information systems, except for studies related to algorithmic models. Furthermore, the well-known Object Oriented Method (OOM) has not been found to be incorporated into current mobile multimodal user interfaces as an architecture that can be used to model them efficiently and reliably.

In this book we will also test the OO methodology by adding multiple mobile modalities into the model while using educational applications as test beds for the interaction of users. The object-oriented model is evaluated for its capabilities to significantly represent the available information from the modalities of interaction, as well as the information about the users and their actions. The resulting model incorporates all common object oriented concepts, which are described explicitly in Sect. 2.3, trying to provide exploitable information, more robustness for the algorithms that are going to process the available data and easiness in the addition of potential new modalities for the purposes of providing even more advanced and adaptive user interfaces.

This book also aims to present and evaluate a prototype mobile learning system with emotion recognition and user personalization capabilities through the OO approach. This effort is quite demanding for two basic reasons. First of all, real time emotion recognition in educational applications is in very early stages of development and as it is stated in Picard (2003), people's expression of emotion is so idiosyncratic and variable, that there is little hope of accurately recognizing an individual's emotional state from the available data. This problem becomes even larger since there are considerable limitations in a system's processing and data storing capabilities such as in mobile or handheld devices. Furthermore, the actual interaction between a user and his/her mobile device is also limited in comparison with the human–computer interaction (smaller keyboard, smaller screen with lower resolution and lower quality in microphones and cameras). In view of the above, this book's approach is to create a system with affective real-time interaction capabilities that works as a remote server and is responsible for recognizing emotional states from the available incoming data. In order to accomplish this, mobile devices get wirelessly connected to the emotion detection server through an object oriented interface and thus, considering the affective interaction, mobile phones are used as input and output devices. This process is comprehensively described in Chaps. 7 and 8.

References

Bernhardt D, Robinson P (2008) Interactive control of music using emotional body expressions. In: Proceedings of conference on human factors in computing systems, pp 3117–3122

Chan YY, Leung CH, Albert KW, Chan SC (2003) MobiLP: a mobile learning platform for enhancing lifewide learning. In: 3rd IEEE international conference on advanced learning technologies, 2003, Athens, Greece, p 457

Cumming G, McDougall A (2000) Mainstreaming AIED into education? Int J Artif Intell Educ 11:197–207

Esposito A (2009) The perceptual and cognitive role of visual and auditory channels in conveying emotional information. Cogn Comput J 1(2):268–278

Goleman D (1995) Emotional intelligenc. Bantam Books, New York

Jascanu N, Jascanu V, Bumbaru S (2008) Toward emotional e-commerce: The customer agent, Lecture notes in computer science (including subseries lecture notes in artifi-cial intelligence and lecture notes in bioinformatics) vol 5177 LNAI, Issue PART 1, pp 202–209

Ketamo H (2003) An adaptive geometry game for handheld devices. Educ Technol Soc 6(1):83–95

Leung CH, Chan YY (2003) Mobile learning: a new paradigm in electronic learning. In: 3rd IEEE international conference on advanced learning technologies, 2003, Athens, Greece, pp 76–80

Murray T (1999) Authoring intelligent tutoring systems: an analysis of the state of the art. Int J Artif Intell Educ 10:98–129

Pantic M, Rothkrantz LJM (2003) Toward an affect-sensitive multimodal human-computer interaction. Proc IEEE 91:1370–1390

Picard RW (2003) Affective computing: challenges. Int J Hum Comput Stud 59(1–2):55–64

Vavoula G, Sharples M (2002) KleOS: A personal, mobile, knowledge and learning organisation system. In: IEEE international workshop on wireless and mobile technologies in education (WMTE'02), pp 152–155

Vincent J (2009) Affiliations, emotion and the mobile phone. Lect Notes Comput Sci 5641:28–41

Virvou M, Alepis E (2004) Mobile versus desktop facilities for an e-learning system: users' perspective. Conf Proc IEEE Int Conf Syst Man Cybern 1:48–52

Virvou M, Alepis E (2005) Mobile educational features in authoring tools for personalised tutoring. Comput Educ 44(1):53–68

Wang CY, Liu B, Chang KE, Horng JT, Chen GD (2003) Using mobile technologies in improving information awareness to promote learning performance. In: 3rd IEEE international conference on advanced learning technologies, 2003, Athens, Greece, pp 106–109, print ISBN: 0-7695-1967-9

Chapter 2
Related Work

Abstract In this chapter we present related work in the scientific area of mobile learning and we also investigate current research efforts in the area of affective mobile computing. Furthermore, we give a brief overview of past works and approaches that have been based on Object-Oriented programming. As one may observe, mobile-learning, in its modern concept, traces its roots to the first decade of 2000. More specifically, a small number of mobile learning applications were built during the first half of the decade with a very high scientific and social interest in this area. Vincent (2009) states that over one half of the world's population is expected to be using mobile phones by 2009 and many people have become attached to and even dependent on mobile devices. In the same study mobile phones are found to maintain close ties within peoples' families and friends. As expected, during the last 5 years there has been a continuously increasing development, using modern technology, both in mobile software and hardware driven mostly by economic benefits.

2.1 Mobile Learning

Computer Assisted Learning (CAL) has grown enormously during the past decades and has been enhanced by the recent advances in web-based applications, multimedia technology, intelligent systems and software engineering. CAL may be used by instructors in a complementary way for their courses. Students may use educational software inside and outside classrooms in order to learn, practice and consolidate their knowledge. They may also use software from remote places in cases where the instructor is far from the student. The quite recent area of Mobile Assisted Language Learning (MALL) has made its appearance during the last decade and is currently widely used to assist in language learning (Virvou et al. 2011). MALL has evolved to support students' language learning with the increased use of mobile technologies such as mobile phones, mobile music players, PDAs and mobile smartphone devices.

E. Alepis and M. Virvou, *Object-Oriented User Interfaces for Personalized Mobile Learning*, Intelligent Systems Reference Library 64,
DOI: 10.1007/978-3-642-53851-3_2,

However, many researchers (Salomon 1990; Welch and Brownell 2000) point out that technology is effective when developers thoughtfully consider the merit and limitations of a particular application while employing effective pedagogical practices to achieve a specific objective. For example, (Hasebrook and Gremm 1999) argue that learning gains are mainly due to instructional methods and thus many researchers aim at making their tutoring systems more effective using "intelligent" software technologies to adapt to the learners' demands, abilities and knowledge. The same applies to web-based educational applications, which are often limited to the capabilities of "electronic books" with little scope for interactivity with the student.

Intelligence can be added to educational software if ITSs technology is used. ITSs have been designed to individualise the educational experience of students according to their level of knowledge and skill (Du Boulay 2000). It has been widely agreed that an ITS should consist of four components, namely the domain knowledge, the student modelling component, the tutoring component and the user interface (Self 1999; Wenger 1987). The domain knowledge consists of a representation of the domain to be taught (e.g. Biology, Chemistry, etc.). The student modeling component involves the construction of a qualitative representation that accounts for student behavior in terms of existing background knowledge about the domain and about students learning the domain (Sison and Shimura 1998). Student modeling may model various aspects of a student, such as what his/her knowledge level is in the domain being taught (e.g. (Sison and Shimura 1998)), whether s/he has misconceptions (e.g. in previous research of our own (Virvou 2002; Virvou and Kabassi 2001)), what the associated emotions with the process of learning are (Elliott et al. 1999) etc. The tutoring component contains a representation of the teaching strategies of the system. Finally the user interface is responsible for translating between the system's internal representation and an interface language understandable by the student.

However the development of such educational programs is a hard task that needs quite a lot of effort from domain and computer experts. One solution to this problem is given by ITS authoring tools, which provide user-friendly environments for human instructors to develop their own ITSs in an easy way. Naturally, the reasoning abilities of the resulting ITSs have to be provided by the authoring tools. Therefore the authoring tools incorporate generic and domain-independent methods that can be customized to the particular tutoring domain of each instructor-author. In this sense authoring tools are harder to develop than ITSs but they provide a high degree of reusability. Therefore they are worth the extra effort.

There have been many research efforts to transfer the technology of ITSs and authoring tools over the Internet. A review (Brusilovsky and Peylo 2003) has shown that all well-known technologies from the areas of ITS have already been re-implemented for the Web. This could be expected since Web-based education needs intelligence to a greater extent than standalone applications. There are two important reasons for this: First, most Web-based educational applications are targeted to a much wider variety of users than any standalone application and thus there is a need for greater personalization; second, students usually work with Web-based educational systems on their own (often from home) and cannot benefit from personalized assistance that an instructor or a peer student can provide in the normal classroom settings (Weber and Brusilovski 2001).

Web-based ITSs and ITS authoring tools can be enhanced significantly by incorporating mobile features in them. Many researchers point out that the basic approach to mobile education should be integrative, combining a variety of devices (mobile and non-mobile) via a variety of transmitting techniques (wired and wireless) (Lehner and Nosekabel 2002; Farooq et al. 2002). There have been quite a lot of primary attempts to incorporate mobile features to this kind of educational technology and the results so far confirm the great potential of this incorporation. As an example, (Uzunboylu et al. 2009) investigated the use of integrating use of mobile technologies, data services, and multimedia messaging systems to increase students' use of mobile technologies and to develop environmental awareness. Data was collected using "usefulness of mobile learning systems" questionnaire from a sample consisting of students.

In particular, Ketamo (2003) reports on an adaptive geometry game for handheld devices that is based on student modeling. Ketamo admits that the system developed was very limited and the observed behavior was defined as very simple. However, an evaluation study that was conducted concerning this system showed that the learning effect was very promising with the handheld platform. A quite different approach is described in the system called KleOS (Vavoula and Sharples 2002) which allows users to organize and manage their learning experiences and resources as a visual timeline. The architecture of KleOS allows it to be used on a number of different platforms including handheld devices. However, unlike Mobile Author, none of the above systems deals with the problem of facilitating the human instructor in the educational software development and maintenance. This problem is dealt with to a certain extent by Moulin et al. (2002). The educational software they have built has addressed both the edition of a mobile lesson content and the management of the students in the field. However, their approach depended heavily on geo-referenced data. Therefore, it was beyond the scope of that system to achieve domain-independence and sufficient generality for the creation of an authoring tool that could provide an environment to instructors for them to create an ITS in any domain they are interested in.

Moving on more recent works considering mobile learning, in (Zhan and Deng 2010) the authors designed a portable multiple-channel message servicing technology for educational reasons where the responding speed and service quality were greatly improved. Chen and Gao (2009) expatiate concepts of mobile learning, realization technology of mobile learning and mobile learning model based on 3G technology and provides four models of mobile learning. The authors of (De Sa and Carrico 2009) present a framework which supports end-users while creating their customized tools for their students. Additionally, the framework comprises means for teachers to include persuasive and motivational mechanisms and hints, promoting student engagement while pursuing their learning activities. The authors describe the framework's architecture, its features, including the supporting guidelines and development process, and detail some of the already developed material and the results that emerged during initial trials and case studies, also stressing their contributions to the field of m-learning.

An interesting review (Frohberg et al. 2009) uses a Mobile Learning framework to evaluate and categorize 102 Mobile Learning projects, and to briefly introduce exemplary projects for each category. All projects were analyzed with the criteria: context, tools, control, communication, subject and objective. Although a significant number of projects have ventured to incorporate the physical context into the learning experience, few projects include a socializing context. Although few projects explicitly discuss the Mobile Learning control issues, one can find all approaches from pure teacher control to learner control. Despite the fact that mobile phones initially started as a communication device, communication and collaboration play a surprisingly small role in Mobile Learning projects. Most Mobile Learning projects support novices, although one might argue that the largest potential is supporting. The authors of (Cavus and Uzunboylu 2009) investigate the effect of mobile learning over the critical thinking skills. The students' critical thinking disposition and the usefulness of mobile learning systems (UMLS) were evaluated. Overall, students' attitudes toward the usefulness of a mobile learning system improved at the end of the experimental study. It was found that after the study, the students' creativity improved significantly. Furthermore, researchers found that outdoor experiences influenced students' attitudes positively. Additionally, results indicate that in this study, working collaboratively and sharing information were built into a group activity.

In the scientific literature of mobile authoring, the authors of (Kuo and Huang 2009) propose an authoring tool named Mobile E-learning Authoring Tool (MEAT) to produce adaptable learning contents and test items. In addition, the visualized course organization tool has also been provided to teachers to organize their courses. All functionalities of the MEAT are designed according to the teachers' feedback and their technological learning needs. To evaluate the MEAT, the authors have conducted an extensive comparison between the MEAT and other (adaptation) content authoring tools. The results indicate that MEAT can produce adaptable contents and test items while supporting learning standard.

The work of (Motiwalla 2007) explores the extension of e-learning into wireless/handheld (W/H) computing devices with the help of a mobile learning (m-learning) framework. This framework provides the requirements to develop m-learning applications that can be used to complement classroom or distance learning. A prototype application was developed to link W/H devices to three course websites. The m-learning applications were pilot-tested for two semesters with a total of 63 students from undergraduate and graduate courses at our university. The students used the m-learning environment with a variety of W/H devices and reported their experiences through a survey and interviews at the end of the semester. The results from this exploratory study provide a better understanding on the role of mobile technology in higher education.

According to the authors of (Wang et al. 2009), acceptance of m-learning by individuals is critical to the successful implementation of m-learning systems and thus, there is a need to research the factors that affect user intention to use m-learning.

The authors investigated the determinants of m-learning acceptance and discovered if there exist either age or gender differences in the acceptance of m-learning. After having tested it, the authors indicated that performance expectancy, effort expectancy, social influence, perceived playfulness, and self-management of learning were all significant determinants of behavioral intention to use m-learning.

2.2 Mobile Multimodal Interaction

In the study of Motiwalla and Qin (2007), the authors explore the integration of speech or voice recognition technologies into m-learning applications. Their voice-enabled mobile application (voice recognition and text-to-speech) not only helps normal users avoid the cumbersome task of typing using small mobile keypads, but also enables people with visual and mobility disabilities to engage in online education. Finally, the results of this study may provide insights into how voice-enabled applications would be received by the blind disabled population. In Feng et al. (2011), the authors highlight the characteristics of mobile search comparing with its desktop counterpart and also provide a review in the state of art technologies of speech-based mobile search.

Current mobile on screen interaction concepts are limited to a two dimensional space with simple or multi-touch capabilities. To this point, the authors of (Hurst and Van Wezel 2011) explore different interaction approaches that rely on multimodal sensor input and aim at providing a richer, more complex, and engaging interaction experience. The authors of this study also conclude that finger tracking interaction seems to be a quite promising approach for mobile gaming and other mobile leisure applications. Spika (2010) has created a software platform based on Java technology for the service logic and XML-based description for the definition of user interfaces that supports services with flexible multimodal user interfaces. Within the fore mentioned platform three basic motion gestures are detected:

- Shaking of the device
- Tilting the device (detection of 6 device orientations)
- Tapping on the device.

The research presented in (Rico 2010) examines the factors affecting social acceptability of multimodal interactions, beginning with gesture-based interfaces. The results of this study revealed the importance of social acceptability, demonstrating that some gestures were significantly more accepted than others. Chittaro (2010) discusses some of the opportunities and challenges of mobile multimodality, which can be a key factor for a better design of mobile interfaces to help people do more on their mobile phones, requiring less time and attention. This study provides important information about interaction differences and similarities between desktop computers and mobile devices and also stresses the need for research towards the effectiveness of different modalities during multiple tasks in mobile conditions. Vainio (2009) concludes that by designing predictive clues and rhythm into mobile multimodal navigation applications, we can improve navigation aids for users. In his paper the author introduced a study that dealt with multimodal navigation and tried to utilize the design practice of episodes of motion that originates from urban planning.

Kondratova (2009) discusses issues associated with improving usability of user interactions with mobile devices in mobile learning applications. The main focus is on using speech recognition and multimodal information such as speech-based interfaces that are capable of interpreting voice commands. The results of this study indicate feasibility of incorporating speech and multimodal interaction in designing applications for mobile devices, while it is also shown that there are considerable limitations in mobile applications including social, technological and environmental factors. The authors of (Reis et al. 2008) have conducted an important study where users had to manipulate a multimodal questionnaire considering different environmental variables, such as lighting, noise, position, movement and type of content. This study investigates the effect of the different context variables in users' choices regarding the available modalities of interaction.

Another example of personalized mobile multimodal interfaces can be found in (Aoidh 2006), where the emphasis is in the personalization of the human–computer interaction with a mobile Geographic Information System (GIS). The author of this paper tries to achieve personalization in the mobile GIS through the implicit and explicit profiling of each user's multimodal, mobile experience. In their paper, Oviatt and Lunsford (2005), highlight three different directions of research that were advancing state-of-the-art mobile technology. These are:

- Development of fusion-based multimodal systems
- Modeling of multimodal communication patterns
- New approaches to adaptive processing.

The authors propose that advances in these research directions may provide more reliable, usable, and commercially promising future mobile systems.

Mueller et al. (2004) introduce an advanced generic Multimodal Interaction and Rendering System that is dedicated for mobile devices, in order to sufficiently integrate advanced multimodal interactions such as mobile speech recognition and mobile speech synthesis. The authors of Shoogle (Williamson et al. 2007) illustrate how model-based interaction can be brought into practical mobile interfaces. Their resulting interface is based around active sensing. According to Oviatt (2000) "One major goal of multimodal system design is to support more robust performance than can be achieved with a unimodal recognition technology". In this paper we may also remark that large gains in decreasing error rates were due to significant levels of mutual disambiguation in the mobile system's multimodal architecture. This study's results confirmed that a more stable multimodal architecture decreased mobile speech recognition error rate by 19–35 %. As Mantyjarvi et al. (2007) claim, Model-based approaches have been recognized as useful for managing the increasing complexity consequent to the many available interaction platforms. Accordingly, the same authors present a development tool as a solution to enable the development of tilt-based hand gesture and graphical modalities for mobile devices in a multimodal user interface.

2.3 Mobile Affective Interaction

2.3.1 Affective Interaction in Computers

Many scientists in the field of affective computing, including Picard (Picard 2003), suggest that one of the major challenges in affective computing is to try to improve the accuracy of recognizing people's emotions. During the last decade, the visual and the audio channel of human–computer interaction were considered as most important in human recognition of affective states (Cowie et al. 2001). At the same time, research in cognitive psychology and in psychophysiology produced firm evidence that affective arousal has a range of somatic and physiological correlates, such as heart rate, skin clamminess, body temperature, etc. (Cacioppo et al. 2000). Apart from verbal interactive signals (spoken words), which are person independent, nonverbal communicative signals like facial expression and vocal intonations are displayed and recognized cross culturally (Pantic and Rothkrantz 2003). A critical point towards the improvement of sophisticated emotion recognition systems is certainly the combination of multiple modalities during human–computer interaction (Caridakis et al. 2010; De Silva et al. 1997; Huang et al. 1998; Oviatt 2003; Zeng et al. 2007). Even less common modalities of interaction are currently investigated, such as in (Busso et al. 2004), where the authors associate features that are derived from the pressure distribution on a chair with affective states. However, progress in emotion recognition based on multiple modalities has been rather slow. Although several approaches have been proposed to recognize human emotions based on facial expressions or speech unimodally, relatively little work has been done in the area of general systems that can use information from multiple modalities that can be added or excluded from these systems in real time.

The problem of effectively combining data from multiple modalities also raises the question of how these modalities may be combined. Correspondingly, this problem consists of the determination of a general architecture of a multi-modal emotion recognition system, as well as of the sophisticated mechanisms that will fuse this system's available data in order to utilize the emotion recognition functions. Classification of recent affect recognizers has been made in (Liao et al. 2006) where affect recognizers have been classified into two groups on the basis of the mathematical tools that these recognizers have used: The first group using traditional classification methods in pattern recognition, including rule-based systems, discriminate analysis, fuzzy rules, case-based and instance-based learning, linear and nonlinear regression, neural networks, Bayesian learning and other learning techniques. The second group of approaches using Hidden Markov

Models, Bayesian networks etc. Indeed, a recent piece of research that uses the above approaches for the integration of audio-visual evidence is reported in (Hwang and Yoon 1981). Specifically, for person-dependent recognition, Zeng and his colleagues (Zeng et al. 2007) apply the voting method to combine the frame-based classification results from both audio and visual channels. For person-independent tests, they apply multi-stream hidden Markov models (HMM) to combine the information from multiple component streams.

In a previous work of ours (Alepis and Virvou 2009) we introduced the OO model within the architecture of a multimodal human–computer interaction system. As a next step, in (Alepis and Virvou 2010) we successfully evaluated this approach by applying it to a desktop affective e-learning application. The results were quite promising and encouraging so as to test this architecture in a more demanding domain that belongs to the very recent research area of affective mobile learning.

After a thorough investigation in the related scientific literature, we come up with the conclusions that there is a shortage of educational systems that incorporate multi-modal emotion recognition capabilities. Even less are the existing affective educational systems with mobile facilities. In (Lim and Aylett 2007) a mobile context-aware intelligent affective guide is described, that guides visitors touring an outdoor attraction. The authors of this system aim mainly at constructing a mobile guide that generates emotions. In our research, we aim at recognizing the users' emotions through their interaction with mobile devices rather than exclusively generating emotions. Another approach is found in (Yoon et al. 2007), where the authors propose a speech emotion recognition agent for mobile communication service. This system tries to recognize five emotional states, namely neutral emotional state, happiness, sadness, anger, and annoyance from the speech captured by a cellular phone in real time and then it calculates the degree of affection such as love, truthfulness, weariness, trick, and friendship. In this approach only data from the mobile device's microphone are taken into consideration, while in our research we investigate a mobile bimodal emotion recognition approach. In (Park et al. 2010), the authors propose and evaluate a new affective interaction technique to support emotional and intimate communication while speaking on the mobile phone. This new technique is called CheekTouch and uses cheeks as a medium of interaction. Tactile feedback is delivered on the cheek while people use multiple fingers as input while holding the mobile phone naturally.

2.3.2 *Affective Interaction in Mobile Devices*

Growing computational power of mobile devices has allowed researchers to make a step further and design applications which "feel what their user feels" (Nielek and Wierzbicki 2010). The authors of this study also state that "Emotion-aware" mobile devices followed by "emotion-aware" services and application are a natural next step in context aware researches. The authors of iFeelng (Rehman and Liu 2010) demonstrate how to turn a mobile phone into a social interface for the blind so that they can sense emotional information of others. In this study we may also find technical details on how to extract emotional information by touch input, as well as how to carry out user evaluations tests in order to test such a system's usability. Finally, the authors conclude that there are channels of communication, such as touch modality, that are not taken into careful consideration yet and which have the potential to enrich mobile phones communication among the users.

The authors of (Rehman et al. 2008) suggest that vibrotactile sensations to sight and sound can make content of mobile phones more realistic and fun. In this study the vibration of a mobile phone is used to provide emotional information. To this end, the possibilities to apply and adopt emotion measuring methods in the field of mobile HCI are investigated in (Geven et al. 2009). In Razak et al. (2008) it is stated that both voice and image are important for people to correctly recognize emotion in telecommunications. The authors of this work have used mobile phones for the interaction of users and their system requires wireless transmission of audio data.

Five emotional states, namely neutral, happiness, sadness, anger, and annoyance recognized by an agent in Cho et al. (2007) through the audio channel of interaction, using mobile phones in real time. As it is stated by the authors of this paper, the available data also contain both speaker environmental noise and network noise which degrade their system's performance in recognizing human emotions from speech. For the classification of the pre-mentioned emotional states k-NN and Fuzzy-SVM approaches have been used and the results are quite promising. Turner et al. (2008) present a survey including 184 young adults is presented in order to explore the relationships between human comfort while making and receiving mobile phone calls in different social contexts and their affective responses to public mobile phone use by others. Yoon et al. (2007) propose the construction of an emotion recognition agent for mobile communication that is based on speech data. Their system tries to recognize users' affect among five emotional states while the experimental results indicate a quite high emotional classification performance (72.5 %).

In this book our prototype systems are incorporated into an educational application and data passes through a linguistic and also a paralinguistic level of analysis in order to address affect. Furthermore, all kinds of multimodal, linguistic and paralinguistic information is stored using the OO model that supports mobile transmission of data during human-mobile device interaction. This proposal is also not found in the related scientific literature.

2.4 Object Oriented Architecture

Object-oriented programming traces its roots to the 1960s, but was not commonly used in mainstream software application development until the early 1990s. Object-oriented design provided researchers with strong frameworks to maintain

Table 2.1 Fundamental concepts in OO modelling and design

OO descriptions	
Class	Classes define abstract collections of object characteristics, including their attributes and their operations
Objects	Objects represent patterns of a class. Objects are created by classes as their templates
Instances	An instance is the actual object created at runtime (program execution)
OO operations	
Operations	Operations are defined as services that can be requested from an object to effect its behaviour. Each operation has a unique signature
Methods	Method is defined as the implementation of an operation. Methods illustrate objects' abilities. In many OO programming languages, methods are referred to as functions
Message passing	Message passing represent the general process by which an object sends data to another object or asks the other object to invoke a method. In general messages define the communication during every kind of object interaction
OO basic concepts	
Inheritance	Subclasses are more specialized versions of a class, which inherit attributes and behaviours from their parent classes, and can introduce their own
Abstraction	Abstraction is simplifying complex reality by modelling classes appropriate to the problem, and working at the most appropriate level of inheritance for a given aspect of the problem
Encapsulation	Encapsulation conceals the functional details of a class from objects that send messages to it
Polymorphism	Polymorphism allows the programmer to treat derived class members just like their parent class' members. Polymorphism in object-oriented programming is the ability of objects belonging to different data types to respond to method calls of methods of the same name, each one according to an appropriate type-specific behaviour
Decoupling	Decoupling allows for the separation of object interactions from classes and inheritance into distinct layers of abstraction
Information hiding	Information hiding represents the restriction of external access to a class's attributes
OO modelling terms	
Aggregation	Represents a relationship between two classes where one class is part of the other class
Composition	Represents a relationship like aggregation but with even "stronger" stronger. In composition a class is treated as a "whole" while another (or other classes) are its constituent "members"
Association	An association represents the way two classes are related to each other
Multiplicity	UML term that models the concepts of cardinality and optionality in an association between two classes
Stereotype (UML)	A stereotype models a common usage of a UML element
Interface	An interface represents a set of definitions of methods and values for which objects agree upon in order to cooperate with each other
Superclass	Superclass is a "mother" class from which other classes are derived
Subclass	Subclass is a "child" class that is derived from another class or classes
Override	Overriding is the action of redefining attributes and/or operations in subclasses

software quality and to develop object oriented applications in part to address common problems by emphasizing discrete, reusable units of programming logic. An object-oriented program may be considered as a collection of cooperating objects, as opposed to the conventional procedural model, in which a program is seen as a list of tasks (subroutines) to perform. In OOP, each object is capable of receiving messages, processing data, and sending messages to other objects and can be viewed as an independent mechanism with distinct roles or responsibilities.

Complementarily to the OO paradigm, the UML (Booch 1996) approach has been developed to standardize the set of notations used by most well-known object oriented concepts. In order to support models deriving by these approaches, Computer Assisted Software Engineering (CASE) tools like Rational Rose (Rational Software Corporation 1997) and Paradigm Plus (Platinum Technology 1997) have been developed.

Table 2.1 illustrates a number of fundamental concepts that are found in the strong majority of definitions of object oriented programming designs.

Object oriented approaches have been already widely used in software development environments (Shieh et al. 1996; Chin et al. 2009). An Object Oriented Learning Activity system is implemented and used to design and perform the learning activity in (Pastor et al. 2001). In (Benz et al. 2004), an object-oriented analysis is adopted for the implementation of remote sensing imagery to GIS. The authors of this paper argue that there is a large gap between theoretically available information and used information to support decision making. As a proposed strategy to bridge this gap, these authors suggest the extension of their signal processing approach for image analysis by the exploration of a hierarchical image object network to represent the strongly linked real-world objects. In these approaches the application of OO architecture has led to several minor or major advantages and has solved many problems. However, the OO approach has not been used for the development of software in the area of affective computing yet.

References

Alepis E, Virvou M (2009) Emotional intelligence in multimodal object oriented user interfaces. Stud Comput Intell 226:349–359

Alepis E, Virvou M (2010) Object oriented architecture for affective multimodal e-learning interfaces. Intell Decis Technol 4(3):171–180

Aoidh EM (2006) Personalised multimodal interfaces for mobile geographic information systems, Lecture notes in computer science (including subseries lecture notes in artificial intelligence and lecture notes in bioinformatics), vol 4018 LNCS, 2006, pp 452–456

Benz UC, Hofmann P, Willhauck G, Lingenfelder I, Heynen M (2004) Multi-resolution, object-oriented fuzzy analysis of remote sensing data for GIS-ready information. ISPRS J Photogrammetry Remote Sens 58(3–4):239–258

Booch G (1996) The unified modeling language. Perform Comput/Unix Rev 14(13):41–48

Brusilovsky P, Peylo C (2003) Adaptive and intelligent web-based educational systems. Int J Artif Intell Educ 13(2–4):159–172

Busso C, Deng Z, Yildirim S, Bulut M, Lee C, Kazemzadeh A, Lee S, Neu-mann U, Narayanan S (2004) Analysis of emotion recognition using facial expressions, speech and multimodal infor-mation. In: Proceedings of the 6th international conference on multimodal interfaces, ACM: State College, PA, USA

Cacioppo JT, Berntson GG, Larsen JT, Poehlmann KM, Ito TA (2000) The Psycho-physiology of emotion. In: Lewis M, Haviland-Jones JM (eds) Handbook of emotions. Guilford Press, NY, pp 173–191

Caridakis G, Karpouzis K, Wallace M, Kessous L, Amir N (2010) Multimodal user's affective state analysis in naturalistic interaction. J Multimodal User Interfaces 3(1):49–66

Cavus N, Uzunboylu H (2009) Improving critical thinking skills in mobile learning. In Procedia—Social and behavioral sciences, vol 1, Issue 1, 2009, pp 434–438

Chen Y, Gao Y (2009) Research on mobile learning based on 3G technology. In: 7th international conference on web-based learning, ICWL 2008, Jinhua, pp 33–36

Chittaro L (2010) Distinctive aspects of mobile interaction and their implications for the design of multimodal interfaces. J Multimodal User Interfaces 3(3):157–165

Cho Y-H, Park K-S, Pak RJ (2007) Speech emotion pattern recognition agent in mobile communication environment using fuzzy-SVM. Adv Soft Comput 40:419–430

Cowie R, Douglas-Cowie E, Tsapatsoulis N, Votsis G, Kollias S, Fellenz W, Taylor J (2001) Emotion recognition in human-computer interaction. IEEE Signal Process Mag 18(1):32–80

De Sá M, Carriço L (2009) Supporting end-user development of personalized mobile learning tools. In: Lecture notes in computer science (including subseries lecture notes in artificial intelligence and lecture notes in bioinformatics), vol 5613, Issue PART 4. LNCS, Portugal, pp 217–225

De Silva L, Miyasato T, Nakatsu R (1997) Facial emotion recognition using multimodal information. In: IEEE international conference on information, communications and signal processing (ICICS'97), pp 397–401

Du Boulay B (2000) Can we learn from ITSs?. In: Gauthier G, Frasson C, VanLehn K (eds) ITS 2000, LNCS 1839. Springer, Berlin, pp 9–17

Elliott C, Rickel J, Lester J (1999) Lifelike pedagogical agents and affective computing: an exploratory synthesis. In: Wooldridge MJ, Veloso M (eds) Artificial intelligence today, LNCS 1600. Springer, Berlin, pp 195–212

Farooq U, Shafer W, Rosson MB, Caroll JM (2002) M-education: bridging the gap of mobile and desktop computing. In: IEEE international workshop on wireless and mobile technologies in education (WMTE'02), pp 91−94

Feng J, Johnston M, Bangalore S (2011) Speech and multimodal interaction in mobile search. In: Proceedings of the 20th international conference companion on world wide Web, pp 293–294

Frohberg D, Göth C, Schwabe G (2009) Mobile learning projects – a critical analysis of the state of the art: original article. J Comput Assist Learn 25(4):307–331

Geven A, Tscheligi M, Noldus L (2009) Measuring mobile emotions: measuring the impossible? In: Proceeding of ACM international conference series. Article number 109

Hasebrook JP, Gremm M (1999) Multimedia for vocational guidance: effects of individualised testing, videos and photography on acceptance and recall. J Educ Multimedia Hypermedia 8(4):377–400

Huang TS, Chen LS, Tao H (1998) Bimodal emotion recognition by man and ma-chine. In: ATR workshop on virtual communication environments, Kyoto, Japan

Hürst W, Van Wezel C (2011) Multimodal interaction concepts for mobile augmented reality applications. Lecture notes in computer science (including subseries lecture notes in artificial intelligence and lecture notes in bioinformatics), vol 6524, Issue PART 2. LNCS, pp 157–167

Hwang CL, Yoon K (1981) Multiple attribute decision making: methods and applications, vol 186. Springer, Berlin

Ketamo H (2003) An adaptive geometry game for handheld devices. Educ Technol Soc 6(1):83–95

Kondratova I (2009) Multimodal interaction for mobile learning. Lecture notes in computer science (including subseries lecture notes in artificial intelligence and lecture notes in bioinformatics), vol 5615, Issue PART 2, LNCS, pp 327–334

Kuo Y-H, Huang Y-M (2009) MEAT: An authoring tool for generating adaptable learning resources. Educ Technol Soc 12(2):51–68

Lehner F, Nosekabel H (2002) The role of mobile devices in E-learning- first experiences with a wireless E-learning environment. In: IEEE international workshop on wireless and mobile technologies in education (WMTE'02), pp 103–106

Liao W, Zhang W, Zhu Z, Ji Q, Gray WD (2006) Toward a decision-theoretic framework for affect recognition and user assistance. Int J Hum Comput Stud 64:847–873

Lim MY, Aylett R (2007) Feel the difference: a guide with attitude!. Lecture notes in computer science, vol 4722. LNCS, pp 317–330

Mäntyjärvi J, Paternò F, Santoro C (2007) Incorporating tilt-based interaction in multimodal user interfaces for mobile devices. Lecture notes in computer science (including subseries lecture notes in artificial intelligence and lecture notes in bioinformatics), vol 4385. LNCS, pp 230–244

Motiwalla LF (2007) Mobile learning: a framework and evaluation. Comput Educ 49(3):581–596

Motiwalla LF, Qin J (2007) Enhancing mobile learning using speech recognition technologies: a case study. In: Conference proceedings of 8th world congress on the management of e-business, WCMeB 2007, Article number 4285317

Moulin C, Giroux S, Pintus A, Sanna R (2002) Mobile lessons using geo-referenced data in e-learning. In: Gouarderes SA, Paraguacu F (eds) Intelligent tutoring systems 2002, vol 2363. LNCS, p 1004

Mueller W, Schaefer R, Bleul S (2004) Interactive multimodal user interfaces for mobile devices. In: Proceedings of the Hawaii international conference on system sciences, vol 37, 2004, Article number STMDI07, pp 4545–4554

Nielek R, Wierzbicki A (2010) Emotion aware mobile application. Lecture notes in computer science (including subseries lecture notes in artificial intelligence and lecture notes in bioinformatics) vol 6422, Issue PART 2. LNAI pp 122–131

Oviatt S (2000) Multimodal system processing in mobile environments, UIST (User Interface Software and Technology). In: Proceedings of the ACM symposium, pp 21–30

Oviatt S (2003) User-modeling and evaluation of multimodal interfaces. In: Proceedings of the IEEE, pp 1457–1468

Oviatt S, Lunsford R (2005) Multimodal interfaces for cell phones and mobile technology. Int J Speech Technol 8(2):127–132

Pantic M, Rothkrantz LJM (2003) Toward an affect-sensitive multimodal human-computer interaction. In: Proceedings of the IEEE, vol 91, pp 1370–1390

Park Y-W, Lim C-Y, Nam T-J (2010) CheekTouch: an affective interaction technique while speaking on the mobile phone. In: Proceedings of conference on human factors in computing systems, pp 3241–3246

Pastor O, Gómez J, Insfrán E, Pelechano V (2001) The OO-Method approach for information systems modeling: from object-oriented conceptual modeling to automated programming. Inf Syst 26(7):507–534

Picard, RW (2003) Affective computing: challenges. Int J Human-Comput Stud 59(1–2):55–64

Platinum Technology Inc, Paradigm Plus: Round-Trip Engineering for JAVA, White Paper (1997)

Rational Software Corporation, Rational Rose User's Manual (1997)

Razak AA, Abidin MIZ, Komiya R (2008) Voice driven emotion recognizer mobile phone: proposal and evaluations. Int J Inf Technol Web Eng 3(1):53–69

Réhman SU, Liu L (2008) Vibrotactile emotions on a mobile phone. In: Proceedings of the 4th international conference on signal image technology and internet based systems, SITIS 2008. Article number 4725810, pp 239–243

Réhman SU, Liu L (2010) iFeeling: vibrotactile rendering of human emotions on mobile phones, lecture notes in computer science (including subseries lecture notes in artificial Intelligence and lecture notes in bioinformatics) vol 5960. LNCS, pp 1–20

Reis T, De Sá M, Carriço L (2008) Multimodal interaction: real context studies on mobile digital artefacts. Lecture notes in computer science (including subseries lecture notes in artificial intelligence and lecture notes in bioinformatics), vol 5270. LNCS, pp 60–69

Rico J (2010) Evaluating the social acceptability of multimodal mobile interactions. In: Proceedings of conference on human factors in computing systems, 2010, pp 2887–2890
Salomon G (1990) Studying the flute and the orchestra: controlled vs. classroom research on computers. Int J Educ Res 14:521–532
Self J (1999) The defining characteristics of intelligent tutoring systems research: itss care Precisely. Int J Artif Intel Educ 10:350–364
Shieh C-K, Mac S-C, Chang T-C, Lai C-M (1996) An object-oriented approach to develop software fault-tolerant mechanisms for parallel programming systems. J Syst Softw 32(3):215–225
Sison R, Shimura M (1998) Student modelling and machine learning. Inter J Artif Intell Educ 9:128–158
Spika M (2010) Synchronizing multimodal user input information in a modular mobile software platform. In: Proceedings of the international symposium on consumer electronics, ISCE 2010, Article number 5522701
Turner M, Love S, Howell M (2008) Understanding emotions experienced when using a mobile phone in public: the social usability of mobile (cellular) telephones. Telematics Inform 25(3):201–215
Uzunboylu H, Cavus N, Ercag E (2009) Using mobile learning to increase environmental awareness. Comput Educ 52(2):381–389
Vainio T (2009) Exploring multimodal navigation aids for mobile users. Lecture notes in computer science (including subseries lecture notes in artificial intelligence and lecture notes in bioinformatics), vol 5726, Issue PART 1, LNCS, pp 853–865
Vavoula G, Sharples M (2002) KleOS: a personal, mobile, knowledge and learning organisation system. In: IEEE international workshop on wireless and mobile technologies in education (WMTE'02), pp 152–155
Vincent J (2009) Affiliations, emotion and the mobile phone, Lect Notes Comput Sci 5641 LNAI:28−41
Virvou M (2002) A cognitive theory in an authoring tool for intelligent Tutoring systems. In: proceedings of IEEE international conference on systems man and cybernetics 2002 (SMC'02), vol 2, pp 410–415
Virvou M, Kabassi K (2001) F-SMILE: an intelligent multi-agent learning environment. In: proceedings of IEEE international conference on advanced learning technologies 2002 (ICALT'02)
Virvou M, Alepis E, Troussas C (2011) MMALL: multilingual mobile-assisted language learning. In: Proceedings of the 1st international symposium on business modeling and software design 2011—BMSD 2011, pp 129–135
Wang Y-S, Wu M-C, Wang H-Y (2009) Investigating the determinants and age and gender differences in the acceptance of mobile learning. Br J Educ Technol 40(1):92–118
Weber G, Brusilovski P (2001) ELM-ART: an adaptive versatile system for web-based instruction. Int J Artif Intell Educ 12:351–384
Welch M, Brownell K (2000) The development and evaluation of a multimedia course on educational collaboration. J Educ Multimedia Hypermedia 9(3):169–194
Wenger E. (1987) Artificial intelligence and tutoring systems. Morgan Kaufmann, Los Altos
Williamson J, Murray-Smith R, Hughes S (2007) Shoogle: excitatory multimodal interaction on mobile devices. In: Proceedings of conference on human factors in computing systems, 2007, pp 121–124
Yoon WJ, Cho YH, Park KS (2007) A study of speech emotion recognition and its application to mobile services. Lecture notes in computer science, vol 4611. LNCS, pp 758–766
Zeng Z, Tu J, Liu M, Huang T, Pianfetti B, Roth D, Levinson S (2007) Audio-visual affect recognition. IEEE Trans Multimedia 9:424–428
Zhan H-F, Deng H–J (2010) Study on a portable education administration assistant system. In: WRI international conference on communications and mobile computing, CMC 2010, vol 1, pp 555–558

Chapter 3
Mobile Student Modeling

Abstract Student modeling is a subset of user modeling, where a student is a specific type of a user who handles a computer system. Focusing on mobile student modeling the student interaction is accomplished through mobile devices' user interfaces. However, the model in its theoretic dimension remains the same. In this chapter the authors give a short presentation of the concepts of user models and user stereotypes. These concepts are used in the following chapters, in all resulting mobile educational systems in order to provide adaptation to the students' personal profiles.

3.1 User Models

From a computer's perspective a "User Model" expresses the computer's need to understand humans and store this information inside its architecture. In computer literature, user model is a term representing a set of personal data associated with a specific user. Usually, a user model contains personal information, such as user profile data. This kind of data is known as explicit information, because most of the times it is data provided by the user himself. However, more recent user models contain users' preferences, habits and even more complicated data that derive from user stereotypes. Implicit information about a user may arise from simple stereotypic data, such as simple "if–then" logic rules. Additionally, more complex implicit information about a user may arise from quite complex processes, where artificial intelligence is used to make a computer "think" as a human. To this end, artificial intelligence algorithmic approaches include Neural Networks (Stathopoulou and Tsihrintzis 2007), Support Vector Machines (Lampropoulos et al. 2011) and most recently Artificial Immune Systems (Sotiropoulos and Tsihrintzis 2011).

In each case, the main goal of constructing such models is providing adaptation and customization of computerized systems to each user's specific needs and capabilities. In this sense, software becomes more efficient and operates in order to

E. Alepis and M. Virvou, *Object-Oriented User Interfaces for Personalized Mobile Learning*, Intelligent Systems Reference Library 64,
DOI: 10.1007/978-3-642-53851-3_3, © Springer-Verlag Berlin Heidelberg 2014

Table 3.1 Categories of user models

User model	Description
Static user models	The first created user models where static, they could not change in time. These models are built in the beginning of human–computer interaction and are not expected to be updated
Non-static user models	These models are dynamic and as a result represent more sophisticated models than the basic static ones. Dynamic user models allow changes and updates in their structure depending on changes in human behavior
Stereotypic user models	Stereotypic user models base their existence in statistics. General user information is gathered, processed and classified in order to build groups of users. As a next step, each new user in a system is assigned to one of the existing groups. Then, as a plausible assumption, this user is expected to act as a common user in his/her group does
Adaptive user models	Adaptive user models are the most sophisticated from the existing user models. These models combine the methods of all three afore mentioned techniques. They try to gather as much information as possible from the user explicitly, they use stereotypic information where the explicit data are insufficient and dynamically change in accordance to the time users spend for their interaction with computers

best serve peoples' needs. Of course, there is a great need in specifying which data is to be included in a user model, supposing there is theoretical superset of all data that can exist in a user model. This need depends mostly on the purpose each software application is created for Table 3.1 illustrates four basic categories of user models that are commonly used in modern software systems.

Correspondingly, we may come up with two basic methods for "mining" data from users in order to build their user models. These methods are illustrated in Table 3.2.

Finally, it is easily comprehensible that a student model is essentially a user model, where a student plays the role of the general user. We may see the student model as a subset of the more abstract user model. To this end, the information that needs to be gathered and processed is about students who usually interact with an educational application.

3.2 User Stereotypes

As already mentioned in the previous subsection, stereotypes constitute a quite powerful mechanism for building user models. This observation was first introduced by (Kay 2000). Their power arises from the fact that stereotypes represent information that enables the system to make a large number of plausible inferences on the basis of a substantially smaller number of observations (Rich 1989, 1999). Stereotypes were first introduced by Rich (1999). Stereotype-based reasoning takes an initial impression of the user and uses this to build a user model based on

Table 3.2 Methods used for user model data mining

Method	Description
Ask the user	Explicit data gathering. We ask users to provide us with information about their personal profile. Such information includes their name, age, gender, skills, interests and knowledge in specific domains. This method is commonly found in user registration
Observe the user	lImplicit data gathering. The system observes the users while they interact with it. The observation is achieved through the available modalities of human–computer interaction. The system "learns" about users by tracking their behavior while using it. Artificial intelligence is most commonly used in this domain

default assumptions. According to Kay (2000), stereotypes consist of the following four basic components:

(a) A set of trigger conditions,
(b) A set of retraction conditions,
(c) A set of stereotypic inferences, and
(d) Threshold probability for inferences.

Trigger conditions are Boolean expressions that activate a specific stereotype. The action of a stereotype is to make large numbers of inferences when a trigger becomes true. The retraction conditions are responsible for deactivating an active stereotype. Once the user is assigned to a stereotype, the stereotype inferences of this particular stereotype serve as default assumptions for the user. Furthermore, threshold probability for inferences is the probability that captures the minimum probability of each inference for a population of users matching this stereotype.

The effectiveness of stereotype reasoning depends on the quality of the identified stereotypes, for example the number of different stereotypes supported by the system, the accuracy of the classification of users to stereotypes, and the quality of inferences that are drawn from stereotype membership. Therefore, before using such an approach, the developers of the ITS should conduct extensive empirical studies to ensure that the supported stereotypes are adequate.

The basic function of a stereotype is to make large numbers of inferences when a trigger becomes true. Once the user is assigned to a stereotype, the stereotype inferences of this particular stereotype serve as default assumptions for the user.

Stereotypes have been used successfully in many educational systems in the past for user model initialization (Bontcheva 2002; Virvou and Moundridou 2001). Yet, in our educational system stereotypes are used for improving the system's emotion recognition accuracy and make the educational interaction affective. Indeed, as Goleman (1995) points out, how people feel may play an important role on their cognitive processes.

Although stereotypes constitute a common user modeling technique for drawing assumptions about users belonging to different groups, it is not very common for stereotypes to be used in conjunction with a multi-criteria decision making theory. In two different cases though, in an e-commerce system called Travel Planner (Chin and Porage 2001) and in a training system called WebIT (Kabassi and Virvou 2004), stereotypes were used in conjunction with a multi-criteria decision making theory. In that system, similarly to the proposed approach, stereotypes were used to draw assumptions about the weights of several attributes. However, the main difference of the proposed approach is that in this case we have multiple stereotypes for describing each user. Furthermore, the use of stereotypes in combination with a multi-criteria decision making theory for affective interaction is a quite different application domain.

Stereotypes (Rich 1989, 1999) are commonly used in user modeling in order to provide default assumptions about users belonging to the same category according to a classification of users that has previously taken place. In the case of Web-IT, default assumptions include the weight of importance of each attribute for a user belonging to a certain age group.

Stereotypes constitute a common user modeling technique for drawing assumptions about users belonging to different groups. However, it is quite common for stereotypes to be used in conjunction with Multi Attribute Decision Making (MADM) theories. In a system called Travel Planner (Chin and Porage 2001), stereotypes were used in conjunction with the Multi-Attribute Utility Theory, which is a MADM theory. In that system, similarly to Web-IT, stereotypes were used to draw assumptions about the weights of several attributes. However, Travel Planner used a different MADM theory from Web-IT. Moreover, the context of that system referred to traveling preferences of remote customers rather than tutoring needs of remote learners of varying ages.

References

Bontcheva K (2002) Adaptivity, adaptability, and reading behaviour: some results from the evaluation of a dynamic hypertext system. In: Proceedings of second international conference on adaptive hypermedia and adaptive web-based systems, lecture notes in computer science, vol 2347. Springer, Berlin, pp 69–78

Chin D, Porage A (2001) Acquiring user preferences for product customization. In: Bauer M, Gmytrasiewicz P, Vassileva J (eds) Proceedings of the 8th international conference on user modeling, UM 2001, Lecture notes in artificial intelligence (LNAI 2109). Springer, Berlin, pp 95–104

Goleman D (1995) Emotional intelligence. Bantam Books, New York

Kabassi K, Virvou M (2004) Personalised adult e-training on com-puter use based on multiple attribute decision making. Interact Comput 16(1):115–132

Kay J (2000) Stereotypes, student models and scrutability. In: Gauthier G, Frasson C, VanLehn K (eds.) Proceedings of the fifth international conference on intelligent tutoring systems, Lecture notes in computer science, vol 1839. Springer, Berlin, pp 19–30

Lampropoulos AS, Lampropoulou PS, Tsihrintzis GA (2011) A movie recommender system based on ensemble of transductive SVM classifiers. In: Proceedings of the international conference on neural computation theory and applications (NCTA 2011), pp 242–247

Rich E (1989) Stereotypes and user modeling. In: Kobsa A, Wahlster W (eds) User models in dialog systems. Springer, Berlin, pp 199–214

Rich E (1999) Users are individuals: individualizing user models. Int J Hum Comput Stud 51:323–338

Sotiropoulos DN, Tsihrintzis GA (2011) Artificial immune system-based classification in class-imbalanced problems. In: IEEE symposium series on computational intelligence (SSCI 2011). IEEE workshop on evolving and adaptive intelligent systems (EAIS 2011) Article no. 5945917, pp 131–138

Stathopoulou I.-O, Tsihrintzis GA (2007) A neural network-based system for face detection in low quality web camera images. In: Proceedings of the international conference on signal processing and multimedia applications (SIGMAP 2007), pp 53–58

Virvou M, Moundridou M (2001) Student and instructor models: two kinds of user model and their interaction in an ITS authoring tool. In: Proceedings of the eighth international conference on user modeling, lecture notes in arti¢cial intelligence, vol 2109. Springer, Berlin, pp. 158–167

Chapter 4
Mobile Authoring in Educational Software

Abstract Towards the incorporation of sophisticated mobile authoring tools in educational software applications, in this chapter, the authors describe an already developed educational platform. This platform extends and improves a mobile authoring tool, developed by the authors, which has been successfully evaluated in the past. This chapter's subsections include an introduction to the topic, an overview of the resulting mobile platform and also an outline of the system's architecture from an Object Oriented view. A sufficient number of snapshots from the operating authoring tool are also available through this chapter's sections in order to illustrate the authoring process.

4.1 Introduction

Mobile Authoring tools allow human instructors to create their own ITS for the domain they are interested in. For this purpose, human instructors have to insert domain data through a user-friendly interface from any computer or mobile device they wish to use. As a next step an authoring tool should also provide the appropriate reasoning mechanisms that are needed for the creation of a complete ITS. In this chapter the authors describe an improved version of an authoring tool that has been developed and presented in the past (Virvou and Alepis 2005). This authoring tool has been re-built for the Android OS and with a more sophisticated and extendable OO architecture. The new resulting authoring tool is targeted on creating personalized educational software and is named Mobile Authoring Tool, or MAT abbreviated.

First we present the Object Oriented structure of MAT by giving a number of UML diagrams that describe the system's structure both from an internal view and also from an external view. We accomplish an internal view by illustrating class diagrams that provide an overview of how the system works. Looking at the systems from the "outside" we observe the actors that use it in UML use-case diagram illustrations. Figures 4.1 and 4.2 illustrate the system's two basic

E. Alepis and M. Virvou, *Object-Oriented User Interfaces for Personalized Mobile Learning*, Intelligent Systems Reference Library 64,
DOI: 10.1007/978-3-642-53851-3_4,

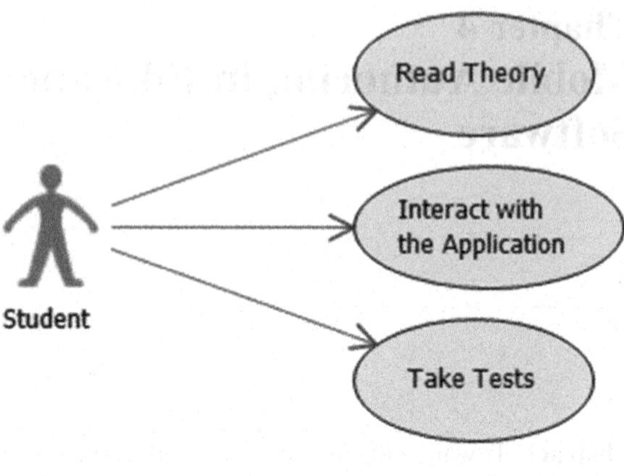

Fig. 4.1 Student's use case diagram

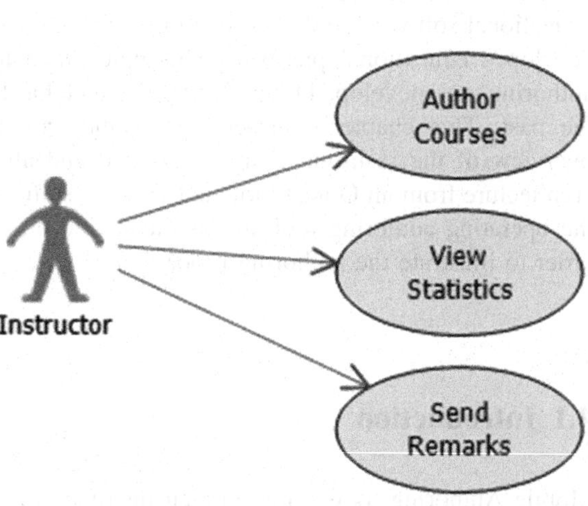

Fig. 4.2 Instructor's use case diagram

operators, or actors, namely the student and the instructor along with their roles in using the system. Additionally, Fig. 4.3 illustrate a more general use case diagram with both students and instructor participating. However, in Fig. 4.3 it is easy noticeable that a general role (of the simple user) is present and both students and instructors inherit use cases from it.

Figure 4.1 illustrates a Use Case Diagram, where the actor is the student. The students have the possibility to read the theory offered by the mobile application, interact with it and take tests in order to evaluate his/her level of knowledge.

Figure 4.2 illustrates a Use Case Diagram, where the actor is the instructor. The instructor has the possibility to author the courses offered to the students, to view the statistics concerning the students and to send remarks to the students.

Figure 4.3 consists of an integrated use case diagram that enlightens the existence of the simple user as a generalization of the actor's student and instructor. The simple user has the possibility to login and logout and to update his personal profile.

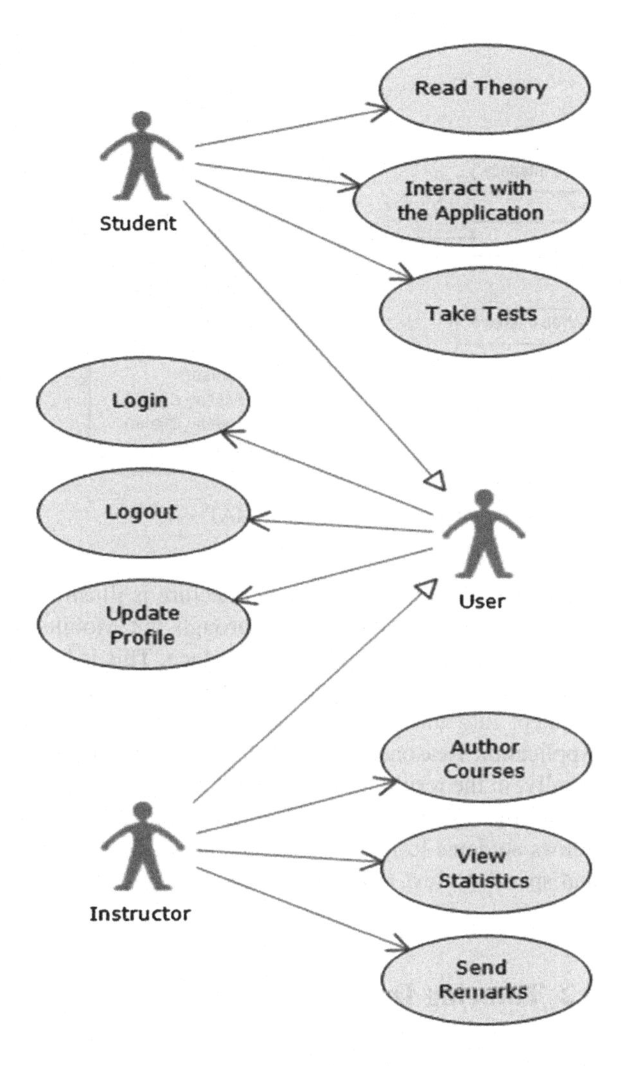

Fig. 4.3 General use case diagram

As a next step, we proceed with an outline of the system's class diagram that concerns the mobile server-client interface (Fig. 4.4). A more complicated class diagram is illustrated in Fig. 4.5, where the first class diagram is improved and explained in higher detail and with more classes incorporated.

Figure 4.5 describes the structure of the mobile application by showing the classes, its attributes, operations/methods, and the relationships among objects. In particular, the class user can choose a Mobile Device through the Device Chooser class and use the educational application, by exploiting the possibilities of his/her mobile device. The educational application has significant features, such as the possibilities of student-instructor communication, the authoring module, the educational reasoning mechanism, which monitors the users' actions while interacting with the application, and the user modeling supporting the stereotypes.

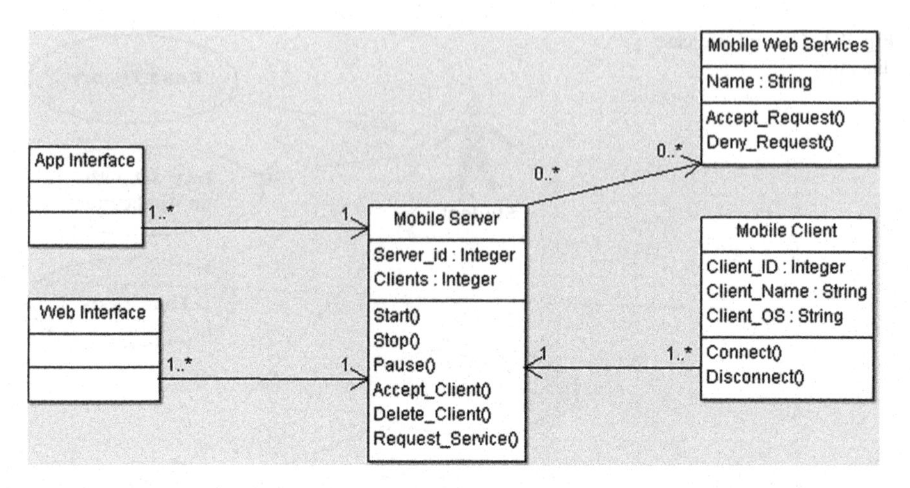

Fig. 4.4 First level class diagram of MAT

A general view of MAT's architecture is illustrated in Fig. 4.6. Instructors can communicate with the system through the Mobile-ITS Server, after they have logged in the application's web interface. This is how they can insert domain data in order to author their curricula. All domain data that are inserted by instructors are kept into the Domain Databases, which communicate with the Educational Application Reasoner and thus can form the ITS to be delivered to the students. Finally, in the resulting mobile application, the students can communicate with the system through a Graphical User Interface (GUI). In addition there is a facility that allows students to interact with the application orally and a text-to-speech (TTS) and speech-to-text (STT) processor can make the appropriate transformations.

4.2 Tutoring Domain

During the authoring procedure, the main task of the tutor-author is to insert domain data into the educational application through the application's user interface. Domain data may consist of lessons that the instructors wish to teach and of student assessment tests, accompanying these lessons. The instructors have the ability to create hypertext documents in the lessons. In the tests, each assessment question may be associated with a part of the theory that students should know so that they may answer the question correctly. Tests may consist of questions of the following types: Multiple choice questions, fill-in the blank space and true/false questions.

Each type of question is associated with certain facilities that MAT may provide to instructors for the creation of a sophisticated educational application. In multiple choice and true/false questions, instructors have the ability to provide a

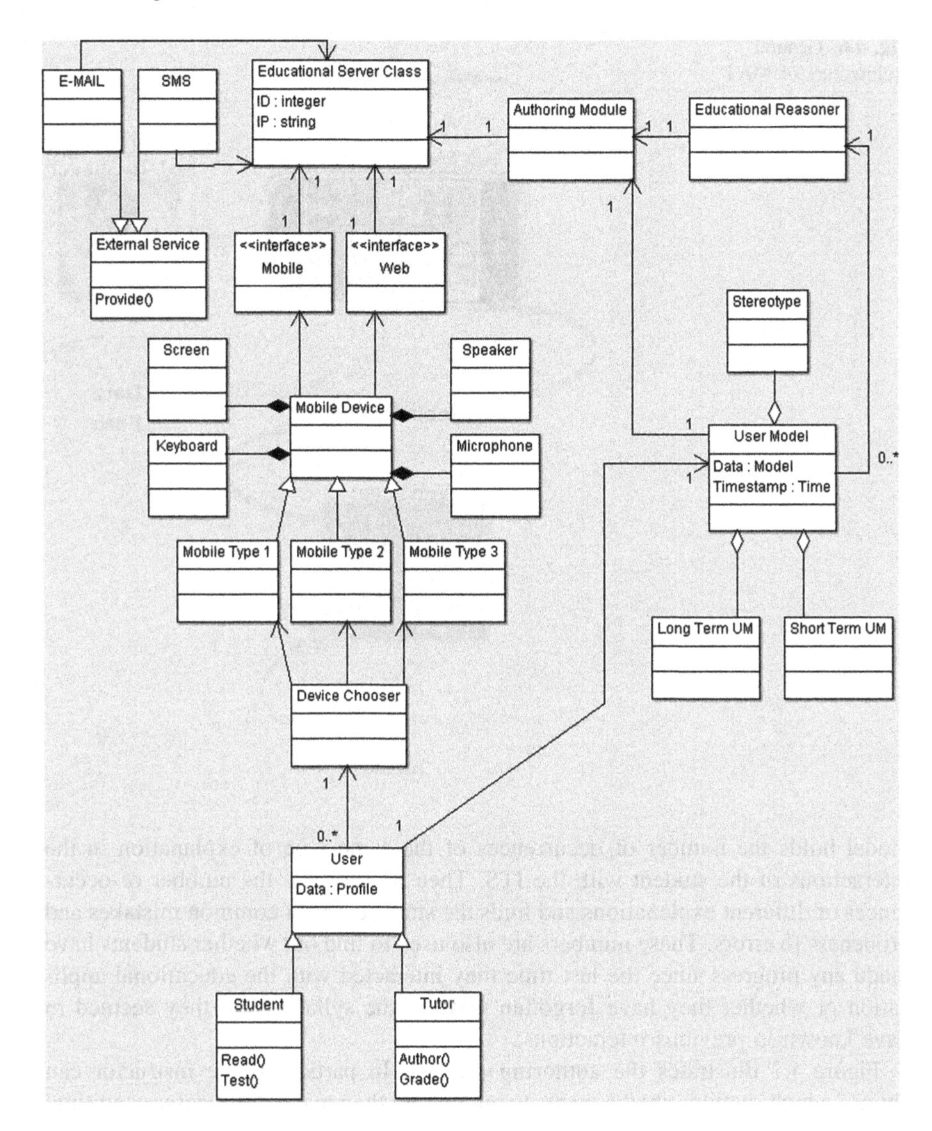

Fig. 4.5 Second level class diagram of MAT

list of frequent errors for each question or they may type explanations of errors. More specifically, they may associate erroneous answers to particular causes and explanations of errors so that these may be used by the system to give more detailed and informative feedback to students. Moreover, these explanations are used to create each student's profile, which is recorded permanently in his/her long term student model and is updated after each interaction of the student with the educational application. For example, the same explanation error may apply to more than one faulty answer of the student. In this case the long term student

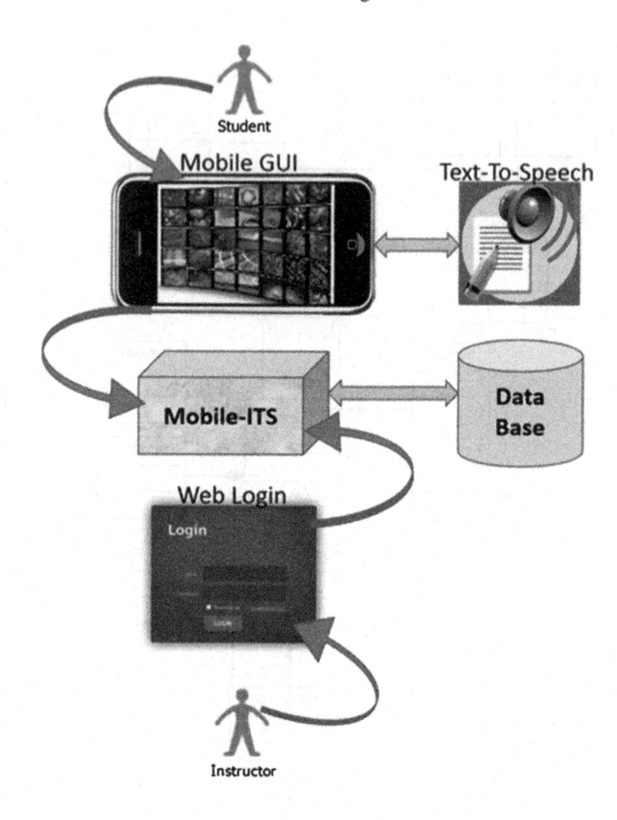

Fig. 4.6 General architecture of MAT

model holds the number of occurrences of the same type of explanation in the interactions of the student with the ITS. Then it compares the number of occurrences of different explanations and finds the student's most common mistakes and proneness to errors. These numbers are also used to find out whether students have made any progress since the last time they interacted with the educational application or whether they have forgotten parts of the syllabus that they seemed to have known in previous interactions.

Figure 4.7 illustrates the authoring process. In particular, the instructor can choose which sections she/he wants to author and then make the necessary editing.

Figure 4.8 illustrates the procedure of authoring a quiz. The instructor can completely change the existing exercise, namely to change the title, the body of the exercise and certainly to mark the right answer.

Figure 4.9 illustrates the procedure of editing quiz settings. Specifically, the instructor can choose when she/he wants to make the quiz available to the students, the duration of the quiz and another information concerning the grading of students.

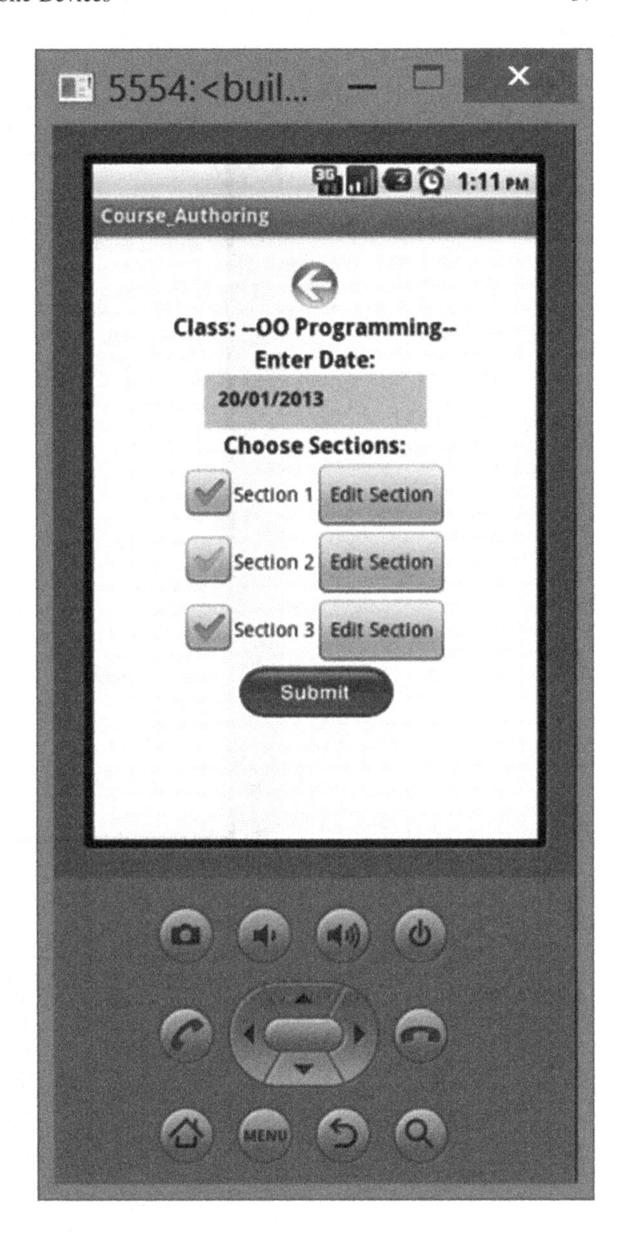

Fig. 4.7 The instructor is authoring a course

4.3 Interconnection with Mobile Devices

The actual way that is used in this chapter for the instructors to communicate with the authoring tool is accomplished through the use of a remote server. This means that they do not need to install any particular application in their mobile device, since the interaction is accomplished through the use of web pages. In particular,

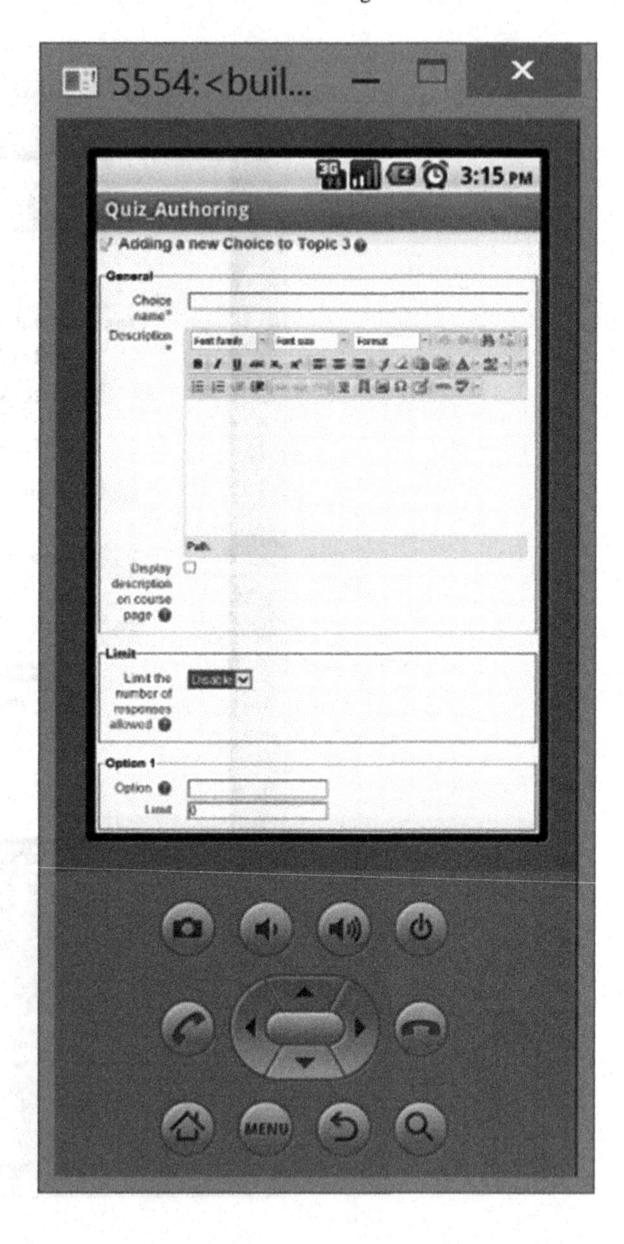

Fig. 4.8 The instructor is authoring a quiz

instructors are able to connect to the system databases through their wireless devices, either mobile phones or mobile Pocket PCs simply by entering the corresponding URL into their devices, which consists of the IP of the server computer and the name of the mobile dynamic web page (example: http://195.252.225.118/mobilelogin.htm). In order to achieve that, we have implemented mobile web pages that use dynamic mobile controls with Java as the programming language.

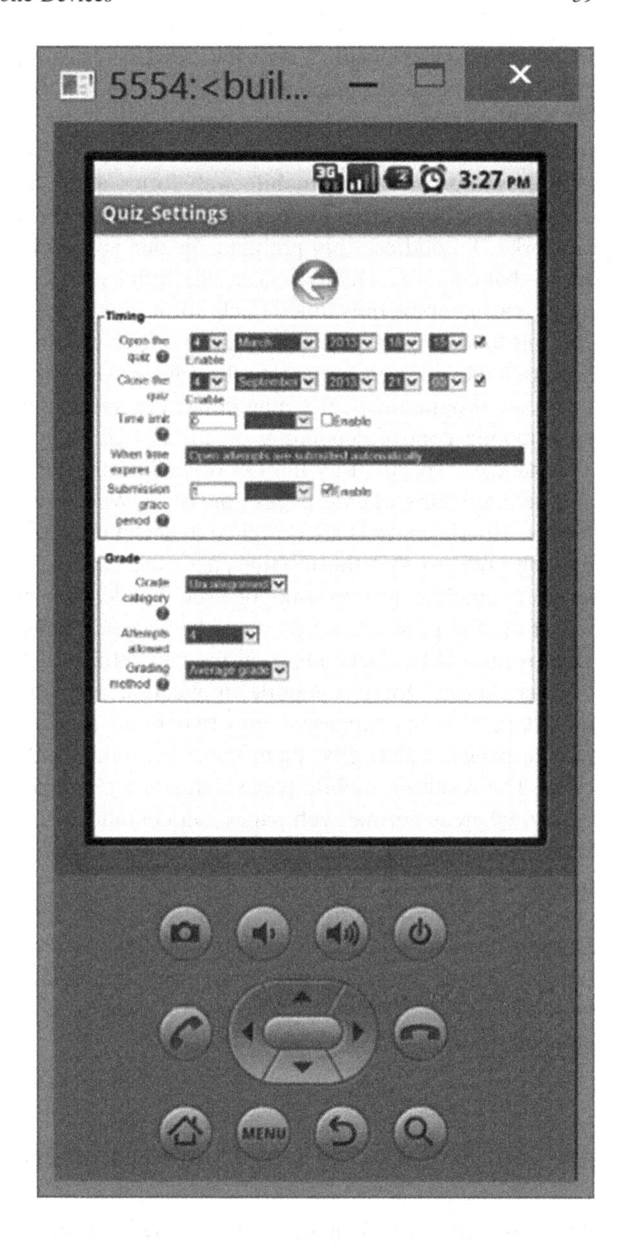

Fig. 4.9 The instructor is editing quiz settings

The server computer of MAT requires use of Windows Server 2008 or higher as the operating system, with a web server pre-installed.

For the development of the mobile forms the Android SDK has been used and more specifically it has been incorporated into the Eclipse Integrated Development Environment (IDE). Any user can install and use the mobile application simply using any device that supports the Android OS although some forms might differ in presentation, depending on each device's hardware specifications. A second requirement is the existence of a wireless mobile network, such as Wi-fi or mobile 3G (or higher) network. This is compulsory in order to connect each mobile client to the main educational server.

Refreshing or loading mobile web forms and even transmitting data through web services often takes quite a long time due to the limited bandwidth of mobile networks. To address this problem, in our approach we have used mobile web forms that contain as many server-side forms as necessary, whereas ordinary web pages can contain only one. Each form can contain different information and controls to display, depending on the end-user's mobile device. This minimizes the need to load new web forms into the device. Moreover our platform's mobile web forms are programmatically manipulated to automatically handle the rendering of each mobile control depending on the device being used. These parameters are automatically specified by the OS of each user's mobile device.

The platform's mobile pages consist of Web forms, server controls and script blocks. Mobile controls are designed to detect the client type and render the correct markup code so that the mobile pages are capable of rendering to a variety of supported mobile devices using one set of code. This provides greater functionality to our mobile pages, since they can be accessed either from a mobile phone or a quite recent tablet. Depending on the capabilities of the devices we have defined "device filters" for our mobile application and for each type of device. It is important also to emphasize into two basic aspects of the architecture of our mobile pages, which give them more flexibility and also reduce network traffic costs. The resulting mobile pages can contain as many server-side forms as necessary, whereas normal web pages can contain only one. This means that mobile controls automatically paginate web content according to the device keeping the paged data on the server until requested by the user. As an example, image controls may allow us to specify multiple image files for one page image, using multiple formats and the correct image file is selected based on device characteristics. These foregoing aspects are very important considering the limited network bandwidth of mobile networks followed by higher costs when compared to computer networks.

4.4 Mobile Tutoring and Course Management

After creating an ITS, it may be used by students as an educational tool while instructors can be assisted in the management of the course and the assessment of their students. As a result, at this stage, both user roles (students and instructors) can use the application to cooperate and/or communicate during the educational process. The instructor and the students should not only be able to have easy access to the data of systems/databases but they should also have an easy way to "communicate" with each other for educational purposes.

The communication between instructors and students can be realized in many ways. By using a mobile phone (and of course connecting to the application's mobile pages) instructors can send short messages via the short message service (SMS), either directly to their students (if they also have mobile phones) or by e-mail. Alternatively instructors can "write" the message to the application's data-base. In this case, instructors have to declare the name of the receiver and the application will use its audio-visual interface to inform him/her as soon as he/she opens the application.

In the first case (e-mail or SMS message) the message is written to the data-base and then is sent to an internet service that provides SMS message and e-mail sending capabilities. Instructors are also able to send an e-mail directly through their mobile phones but it may be preferable for them to use the application to do that. The main reason for this is the fact that mobile networks are considerably slow and cost much and thus they may not be very convenient for the application's users. Thus the application is expected to send an e-mail to the internet service. The body of this e-mail will be the body of the short message and the "subject" of the e-mail will specify the receiver by his/her mobile telephone number. Messages (if sent from the instructor) can contain information about which test the student should visit next or about anything that the instructor thinks that the student should pay attention to. Messages can also be sent by students who can very easily send reports of their progress to their instructors.

Instructors have easy access to the "master" data-base of the application through their mobile phone, which means that they can stay informed of the progress of their students wherever they may be and whenever they wish. Naturally, students also have the "privilege" to access the data-base by their mobile phone. Students would access the data-base for different reasons from instructors. For example, if they wish to see which test is next, or to read the remarks of their instructor about previous tests.

Additionally both instructors and students are able to send short message services (SMSs) containing remarks and additional information. The body of the SMS is entered in the "enter message" field and the name of the receiver is written in the "enter student name" field. Finally after the "direct SMS or Write to Database" choice, the message is sent by pressing the "Send!" button. If the user selects "Direct SMS" then the SMS is delivered directly to the mobile device of the receiver through an Internet service described earlier. Only the username is required since all the mobile phone numbers are stored in the system's database.

Figure 4.10 illustrates the possibility of the instructor to check the students' statistics for the course that she/he is teaching. In this way, the instructor can adapt the curriculum and its level of difficulty based on the specific needs of the students.

The discrimination between instructors and students is conducted by the application (installed in the server) and for each different user a personal profile is created and stored in the database. In order to accomplish this, a user name and

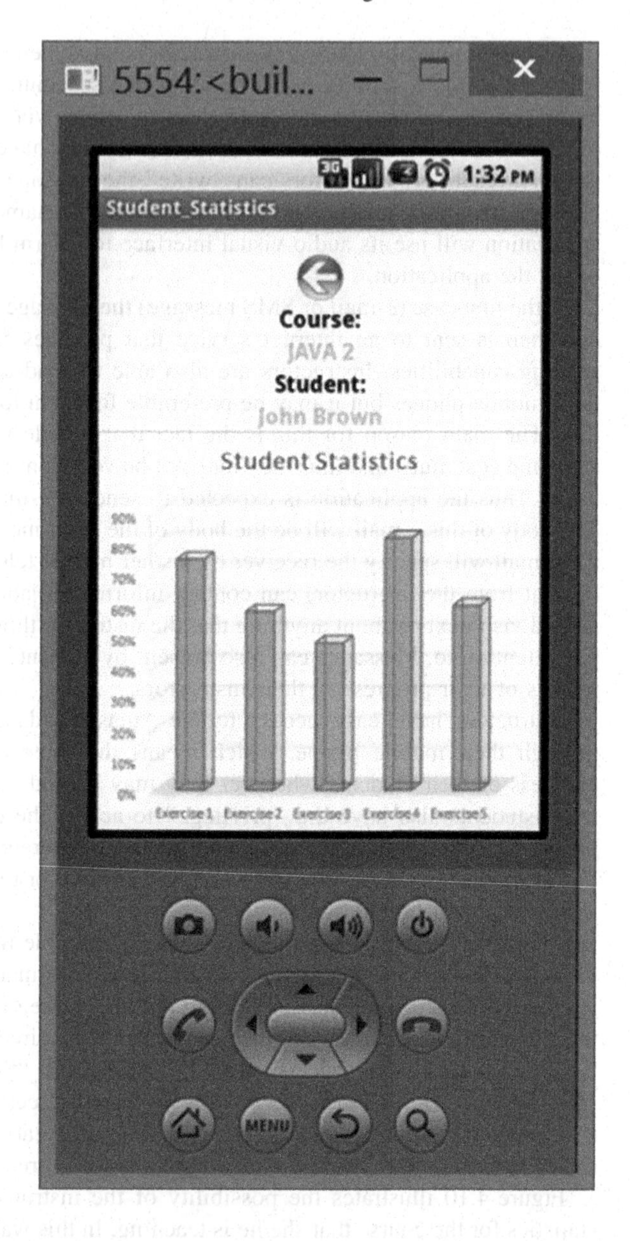

Fig. 4.10 Example: An instructor looks at a student's course statistics

password is always required to gain access to the mobile web pages. On the first level of authentication the application determines whether there is a valid user-name-password or not. On the second level of authentication, users are identified as simple users (students) or supervisors (instructors).

Mentioning the security and privacy aspects of our mobile application, it is also possible to use the Secure Sockets Layer (SSL) in order to encrypt data during important data transactions. Both the system's server and the programming framework for the targeted mobile operating system support this operation simply by inserting a valid certificate signed by a certificate verifying authority. The user can easily use this secure way of communication by inserting an "s" in his/her mobile browser's URL (e.g. https://195.252.225.118/mobilepage.html).

Figure 4.11 illustrates the possible ways of communication between the student and the instructor. This can be achieved via three different ways, namely writing to the Database, SMS service and Application popup.

As mentioned earlier, the ITS that has resulted from the authoring process, uses the Educational Application Reasoner to reason about the students' way of learning. Then all long-term characteristics inferred about a particular student are stored in his/her long term student model so that they can be used by the application itself to be more adaptive to the student's needs or by the instructor so that she/he can provide tutoring advice tailored to each individual student. Thus, the educational application monitors the students' actions while they work with the system. It maintains a record of how many times (if any) a student has visited a particular web page of the theory and simultaneously keeps a record about how the student performs in tests. Then all the information that has been gathered is used by the system so that a personal student profile is created and his/her progress is monitored.

Students may take tests through their mobile phones. An example of a student's examination process is illustrated in Figs. 4.12 and 4.13, where a student first selects a course and as a second step selects the type of quiz she/he would like to use during the examination. When students answer questions, the tutoring system tries to perform error diagnosis in cases where the students' answers have been incorrect. Error diagnosis aims at giving an explanation about a student's mistake taking into account the history record of the student and the particular circumstances where the error has occurred. Giving a correct explanation of a mistake can be a difficult task for a tutoring system. One problem that further complicates this task is the ambiguity, since there may be different explanations of observed incorrect users' actions. For example, in a fill-in-the-blank-space question a student may give an incorrect answer simply because she/he has mistyped the answer. However, this may well appear as a lack of knowledge in the domain of knowledge that is being taught.

MAT always checks the students' answers with respect to spelling or typing errors. In particular, spelling mistakes may be identified by the following procedure. Each time that an incorrect answer is typed by a student, this answer is converted to the actual pronunciation sound of the word typed and is then compared to the pronunciation sound of the correct answer that the student should have given. If these answers are found similar then perhaps the student has made a spelling mistake. If this is the case, it does not mean that she/he does not know the answer at all. If the student has typed a word, which is completely different from the correct one then she/he has made a domain error. Such information about the

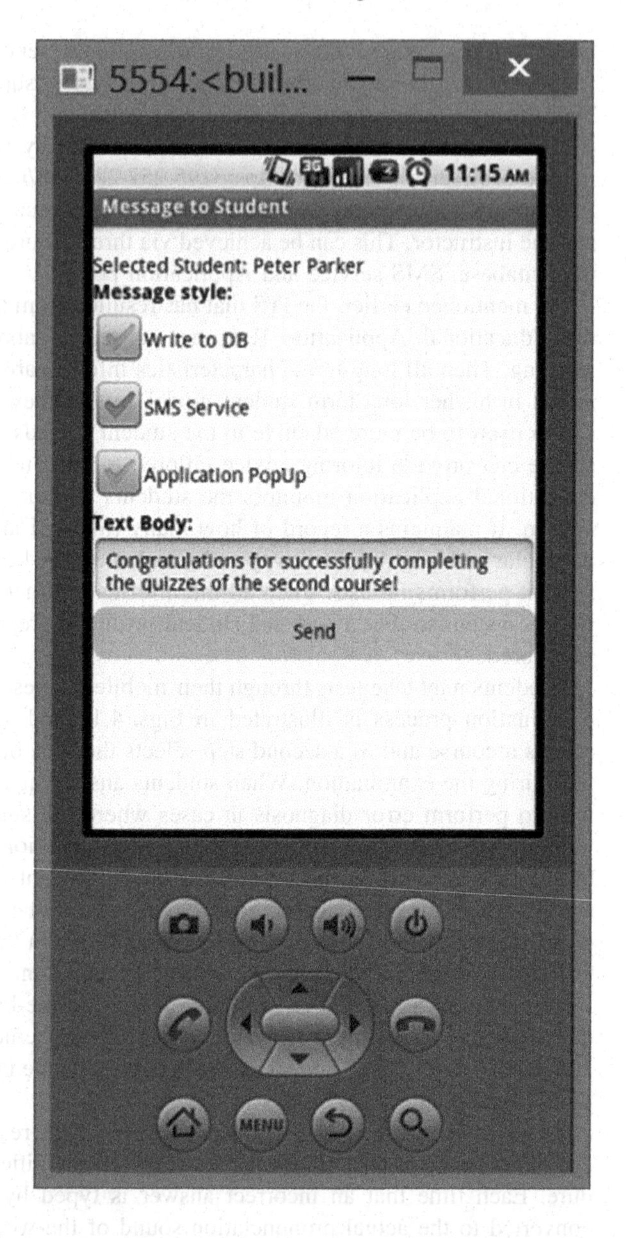

Fig. 4.11 Instructor communicating with student

cause of an error is then recorded to the student model. Thus, the tutoring application records both domain-dependent and domain-independent information about particular students to their long-term individual student models. For example, a student may be consistently making a lot of spelling mistakes when she/he is typing answers to questions posed by the tutoring application. This is a domain-

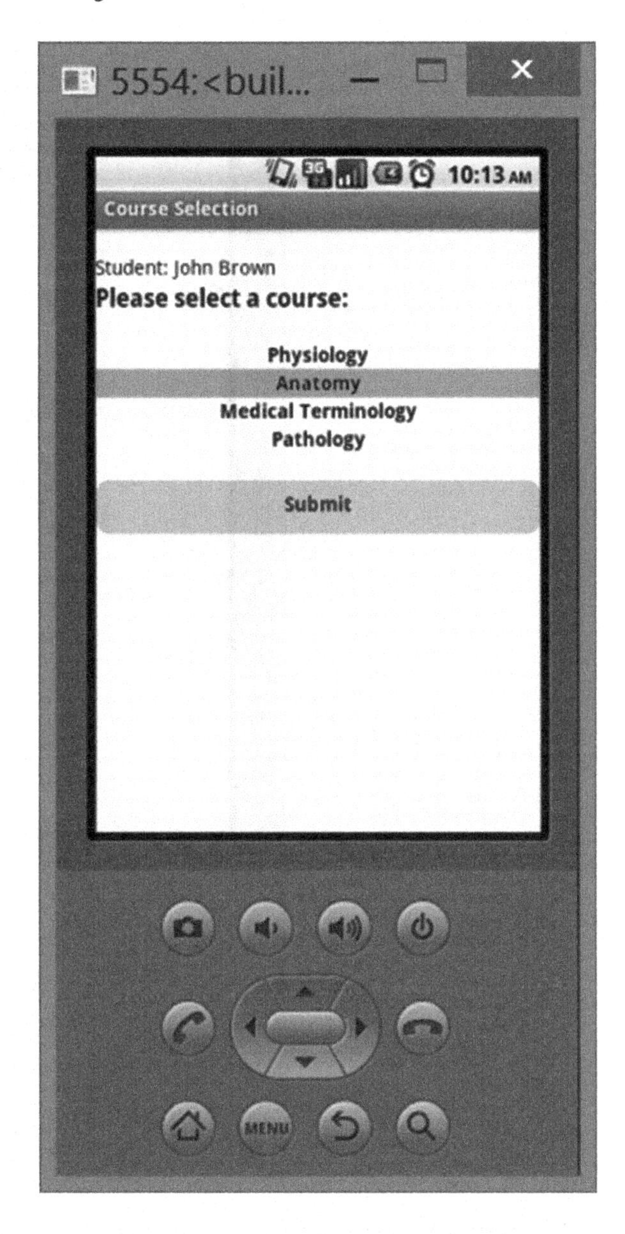

Fig. 4.12 A student is selecting a course to be tested

independent feature of the student concerning the student's spelling skills rather than domain knowledge.

Domain errors may be further examined for the identification of a deeper cause of their appearance. For example, the instructor may have provided a list of frequent errors and each of them may have been associated with an underlying cause of error. In this way the instructors may create a bug-list, which is based on

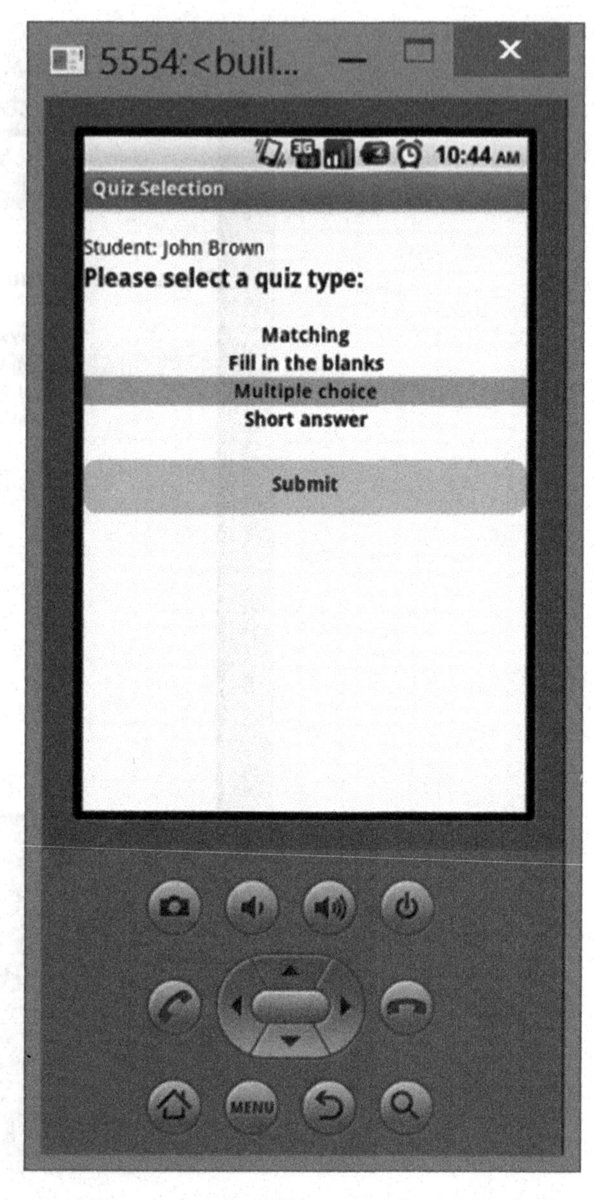

Fig. 4.13 A student is selecting a quiz type to start the examination

their experience about students' making errors. Such lists may be used for further classification of domain errors and the student model is updated.

Reference

Virvou M, Alepis E (2005) Mobile educational features in authoring tools for personalised tutoring. Comput Educ 44(1):53–68

Chapter 5
Extending Mobile Personalization to Students with Special Needs

Abstract This chapter focuses on describing an object oriented architecture that targets on extending mobile educational facilities to students with special needs. This is a common issue in mainstream schools, where students with special needs usually have problems in physical and/or mental participation in classes. These students often need a higher level of supervision and coordination by people related to them such as tutors, parents, therapists and co-students. Mobile computing can offer great opportunities in many cases, as in remote learning, communication, participation and naturally, supervision coordination. This chapter concludes presenting a mobile educational platform that keeps history models of students and records common problems, weaknesses and progress so that it may be used effectively by all parties involved in their education.

5.1 Introduction

Social communities around the world have acknowledged the urging need to give students with special needs equal opportunities to education. Over the last two decades, several countries have led the effort to implement policies which foster integration of students with special needs (Avramidis et al. 2000). The rather daunting challenge for educators and researchers in the field of learning disabilities is to examine effective instructional approaches relative to specific learning disabilities while at the same time recognizing the powerful influence of the social context (Kracker 2000).

However, such challenges often make it difficult for tutors to have students with disabilities in their classes. For example, in a study conducted among tutors of mainstream schools in Hong Kong (Pearson et al. 2003), it was found that most of the tutors (70 %) were in favour of the realisation of equal opportunities for students with disabilities. However, at the same time, many of the tutors who participated in the study, tended to agree with the statements "integration was a burden to the schools and tutors" (over 60 %) and a "painful struggle for special

E. Alepis and M. Virvou, *Object-Oriented User Interfaces for Personalized Mobile Learning*, Intelligent Systems Reference Library 64,
DOI: 10.1007/978-3-642-53851-3_5, © Springer-Verlag Berlin Heidelberg 2014

students" (over 48 %). This is probably the reason why the problems remain unsolved to a large degree in most countries. For example, (Hasselbring and Glaser 2000) from a report, millions of students across the United States cannot benefit fully from a traditional educational program as they have a disability that impairs their ability to participate in a typical classroom environment. They continue to make the point that computer-based technologies can play an especially important role for such students.

Certainly, computers can provide the medium for more personalised learning adapted to the special needs of students with disabilities. Yet the benefits of computer-based learning for students with special needs can be increased significantly through web and mobile technology. There are several reasons for this. First, mobile phones are more widespread than computers, thus they provide wider access to computer programs without the extra cost of buying a computer. This is a significant asset for the educational facilities of students with disabilities. Indeed, as (Hasselbring and Glaser 2000) point out, the use of computer technology can considerably help students with disabilities to keep up with their non-disabled peers; however, the cost of the technology is a serious consideration for all schools. Second, mobile phones provide the ultimate mobility which is so important for students with disabilities and the people who support them (parents, tutors, therapists, co-students). Students with special needs often find it more difficult to be physically present in a class or computer lab. In this respect, the mobile phone can give them access to all computer facilities from their home. Similarly, the people who support students with special needs can have access to educationally assisting computer facilities through their mobile phones too. Thus they also do not need special computer equipment or be at a specific place to physically meet each other about the progress of the students with special needs. This is an extremely important asset for the efficient education of students with special needs. Indeed, as (Barrett 2000) highlights, coordination with agencies, health care providers and families is essential for the education of students with special care needs. Mobile phones can facilitate and even encourage such coordination to a greater extent than other technology equipment.

In Sullivan et al. (2010) it is stated that the adoption of emergency notification systems on university campuses is increasing, providing students and staff the ability to receive emergency notifications though a variety of means, including text and voice messages sent to mobile phones. The authors of this paper are conducting studies in order to show how to better adapt emergency communications using mobile phones to populations with special needs. The study described in (Barbeau et al. 2010) describes an application called TAD (travel assistance device) that aids transit riders with special needs in using public transportation. TAD provides riders with special needs customized real-time audio, visual and tactile prompts in order to increase their level of independence and their caretakers' level of security. For children who cannot verbally communicate, augmentative visual communications tools can enable them to get their needs met, to socialize, and more (Monibi and Hayes 2008). These authors have also created

"Mocotos", a set of mobile tools for the communications of children with special needs.

While much work has been and is being done regarding the opportunities and challenges arising from these technologies, much less exists on the unique opportunities and implications the same devices present and raise to users with special needs (Bertini and Kimani 2003). The same study also gives an overview on the opportunities that mobile devices present to users with special needs such as the disabled, the elderly, and the sick people. Evaluation results shown in (Brooks et al. 2002; Bertini and Kimani 2003) indicated overall client satisfaction with mobile dental unit services in the absence of competent community based dental care for children with special healthcare needs. Participants in this study reported both their gratitude and their frustration that the mobile dental program was the only dental care available to this population.

In view of the above, in this chapter, we describe a mobile learning platform that is meant to help the education of learners with special needs. Thus, it takes into account what these needs are so that it may incorporate as many features as possible that will help the educational process of students with disabilities. The educational process involves both the students themselves and also the social and medical environment consisting of people that supports them.

5.2 Overview of the Mobile Educational Platform

This mobile platform has been initially developed as a pilot study for the needs of three categories of students with special needs, namely students with moving difficulties, sight problems and dyslexia. To achieve this, the mobile platform incorporates the following main features: It incorporates a multi-modal interface so that it may facilitate the interaction with users having disabilities. Users may use a microphone as an alternative way of interaction, rather than the traditional use of keyboard and mouse. The whole interaction with the educational platform is additionally supported through a mobile interface. Figure 5.1 illustrates a mobile connection with an educational platform where a student may use his/her mobile device in order to participate in a test concerning a specific lesson. Secondly, there is a student modelling mechanism that reasons about users' actions and keeps long term information about them so that the interaction is dynamically adapted to their needs. A student modelling component reasons both about the cognitive and the affective state of students as this is revealed by the students' actions during their interaction with the platform. All the above use cases are modeled, using user modelling techniques in two stages. User modelling occurs firstly by considering the special need that the patient has and secondly by the personal data stored for each user. Finally, the platform facilitates the communication of the involved supporting people through communication services that rely either on the educational platform or on specific telecommunications channels.

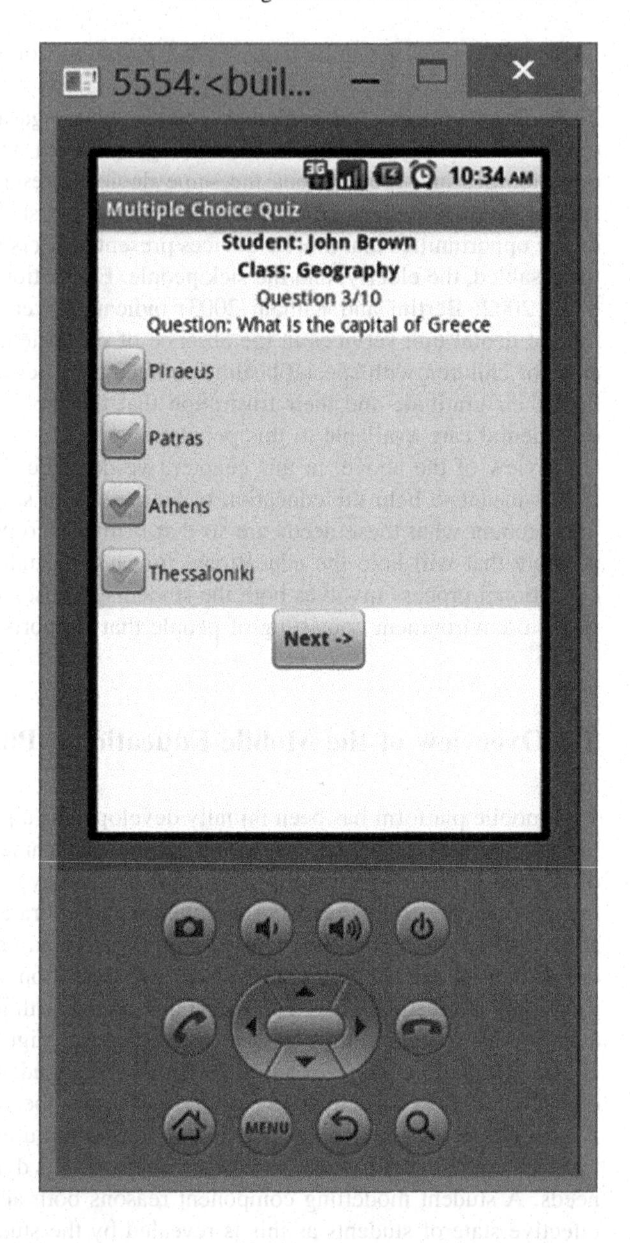

Fig. 5.1 Mobile interaction with the educational platform

The mobile platform can be used by students with disabilities including sight problems, kinetic difficulties and reading disorders. In many cases the interaction can be alternatively achieved visually and/or acoustically. Users can watch the mobile screen and also listen through the use of speech engines that produce synthesized human voices. Users are also able to give oral input to the platform through the use of special software for voice recognition. The commands that are included in the platform as well as the insertion of text can be done orally, through a common microphone. The platform has the potential of being trained and improved according to the user profile that includes characteristics of the user's voice.

Dyslexic students may also benefit from the mobile platform. Dyslexia is a brain malfunction that affects the way that the individual learns, which appears as a difficulty in reading and writing but also a difficulty in obtaining knowledge and skills in other areas of learning as well. Figure 5.2 illustrates the settings in a user's profile, where one or more of the supported disabilities can be selected. For every chosen disability the mobile platform adapts its interface and behavior in order to provide maximum help and friendly interaction to each specific student with special needs.

An outline of the system's UML Class Diagram is illustrated in Fig. 5.3. This class diagram contains basic member classes, a number of associations and multiplicities between them. The only actor that participates as a class in this specific diagram is the user with special needs.

In the following subsections the functionality of the platform is described in terms of the fore mentioned categories of students with special needs.

5.2.1 Students with Moving Difficulties

The platform can interact with people that have kinetic difficulties in their upper parts/ends and are not able to use the keyboard or/even the mouse. In that case the interaction can be done visually and acoustically. The user can see through the use of the screen and also listen through the use of speakers. Correspondingly, the user is able to transfer data (users input) with help from the use of special software for process and voice recognition. The commands included in the platform's educational process (such as the pressing of a specific button command), as well as the typing of plain text, can be achieved orally, alternatively, through the use of a microphone. Figures 5.4 and 5.5 illustrate this kind of interaction. In Fig. 5.4 the mobile application is on hold to hear audio input from a user. Figure 5.5 illustrates the mobile system's voice recognizer that is trying to interact orally with a user.

In each case, the platform has the potential of being trained and improved according to the user profile, with the proper user model that includes characteristics of the user's own voice. The percentage of success increases in accordance with the time the users interact with the platform.

People with moving difficulties (ex. Lower ends) have a reduced freedom of movement and transportation. Thus, we may suppose that the transportation of those individuals to special places equipped with personal computers equipped with the appropriate software would be difficult and in some cases even impossible. Under these circumstances, the use of a computer in a private place (such as the user's own house) is preferable. However, this presupposes that the person that will be supervising (the patient's doctor for example) the individual with special

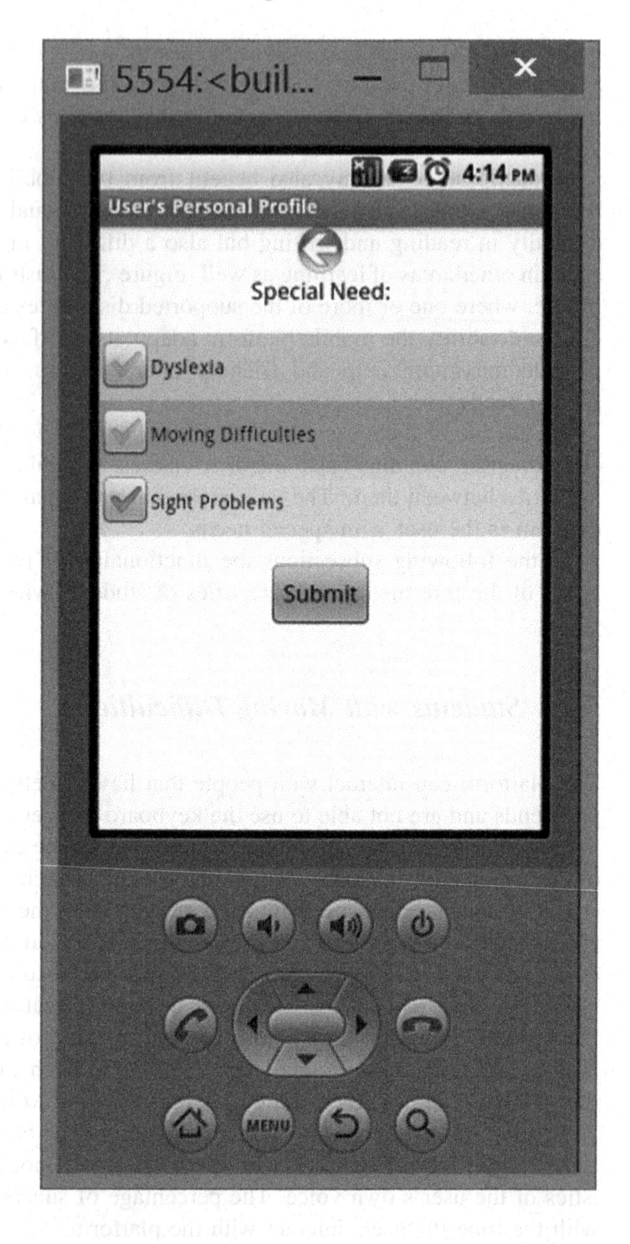

Fig. 5.2 Disabilities settings for each user's profile

needs would be able to move to different places in short periods of time and visit patients, so that he/she can check and evaluate their progress. Evidently, something like this (for example a doctor that would supervise more than 10 patients) would be particularly difficult, if we take into consideration a doctors' extremely heavy and stressful schedule. Such special situations are "resolved" by the platform's basic functionality which is mobile access. With no obligation of buying

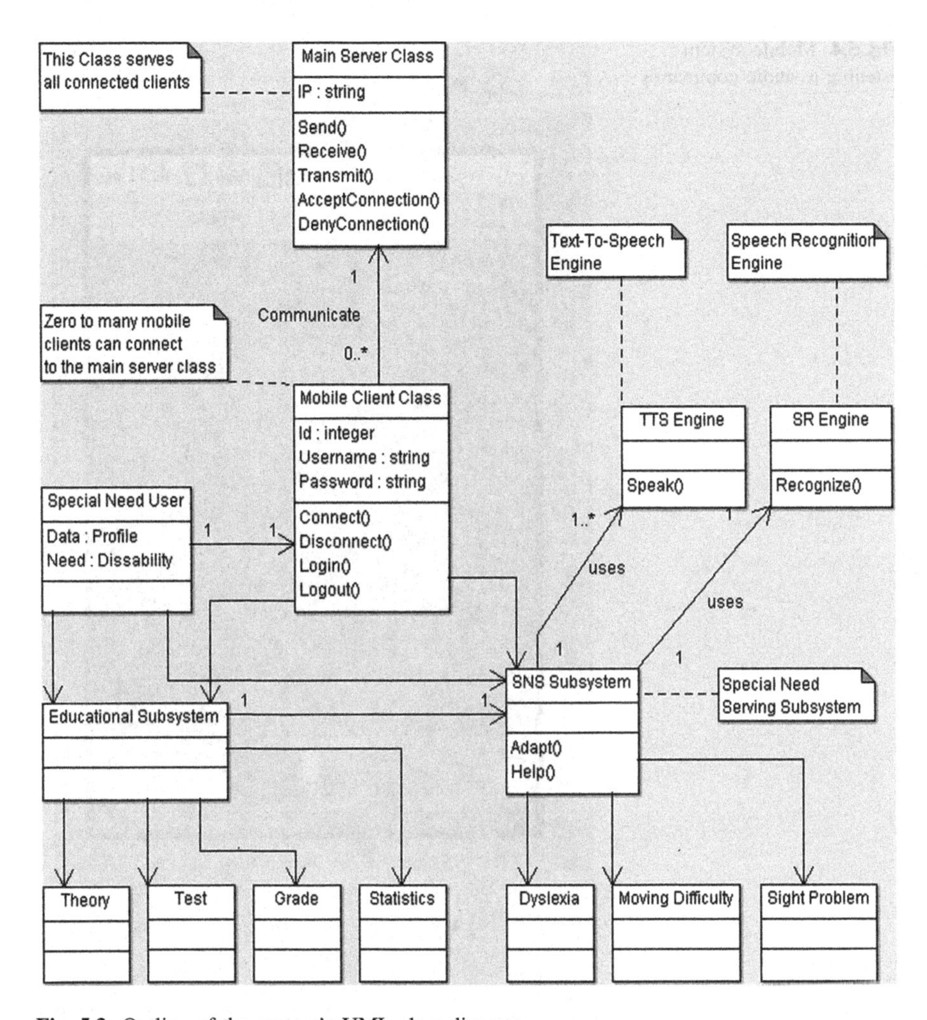

Fig. 5.3 Outline of the system's UML class diagram

extra equipment, doctors may use their mobile phone; connect cordlessly with the internet and gain access to the personal computers of their patients. They can accomplish that anytime, anywhere, thanks to the use of their mobile phone and through an existing web connection. Doctors can also check the progress of their patients, send and receive messages to or from them, and also exchange information and data with the medical platform in reference of each user/patient.

Finally, another important facility provided by the educational platform is its incorporated location service. Through this module, each mobile device, equipped with a Global Positioning System (GPS) sensor is capable of recording and transmitting each user's location. This may be quite important for patients who wish to inform others about their geographical location on a map and also for

Fig. 5.4 Mobile system listening to audio commands

supervising doctors who can easily locate their patients and thus schedule easily their visits to them. Figure 5.6 illustrates the mobile application's location service, where a patient is spotted on the map. Such data can be recorded for future use by third parties, or can be shared with all people that interact with the mobile platform.

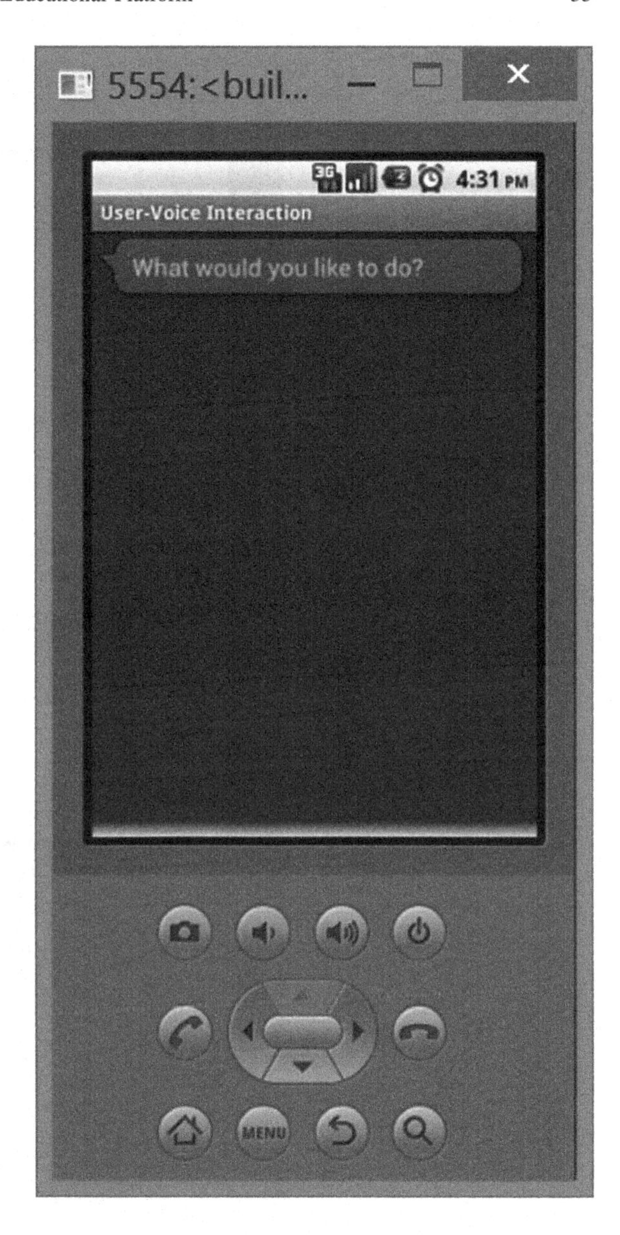

Fig. 5.5 User-mobile phone vocal interaction

5.2.2 Students with Sight Problems

In cases of students with sight difficulties, the interaction may be alternatively acoustical; meaning that the platform may interact with the interacting user by using incorporated speech engines. Additionally, the user is able to communicate orally with the platform through the use of a microphone. Processing and

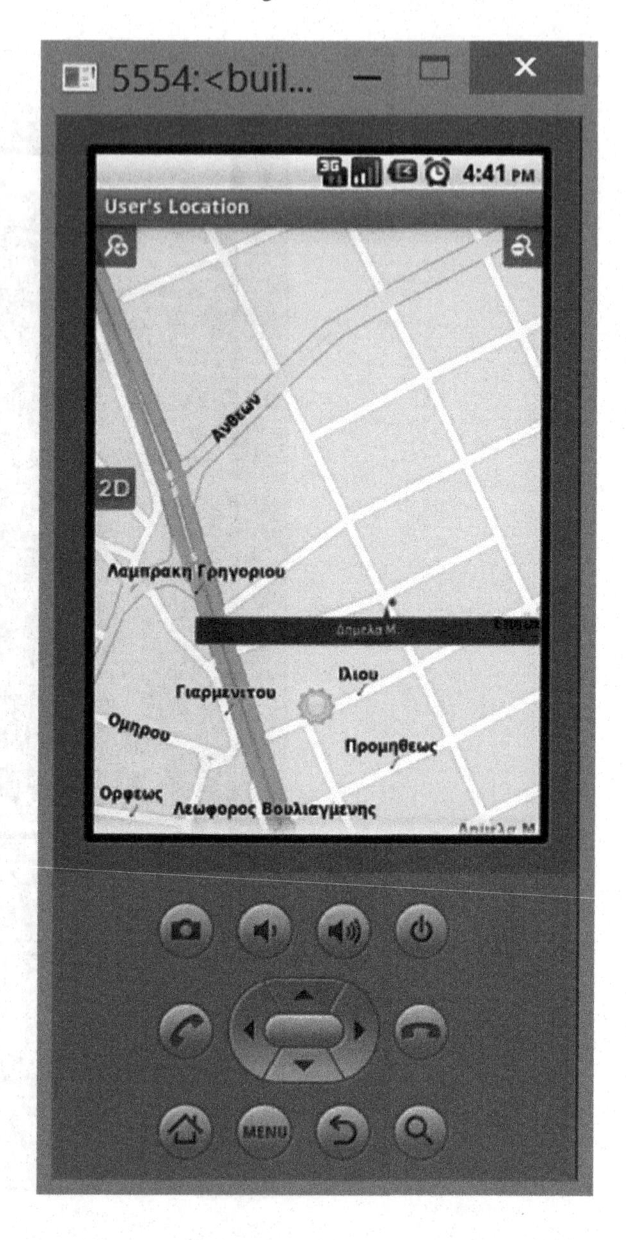

Fig. 5.6 A patient is located on the map

recognition of oral speech take place as a next step. At the same time the user can use a conventional keyboard or a keyboard specially adjusted for people with sight difficulties. Even the use of the mouse is partly evitable, since the movements of the mouse are accompanied with a corresponding vocal reply from the platform. For example, the platform can inform the user at any time for the mousse's actual position, as well as which are the available user actions. Figure 5.7 illustrates an

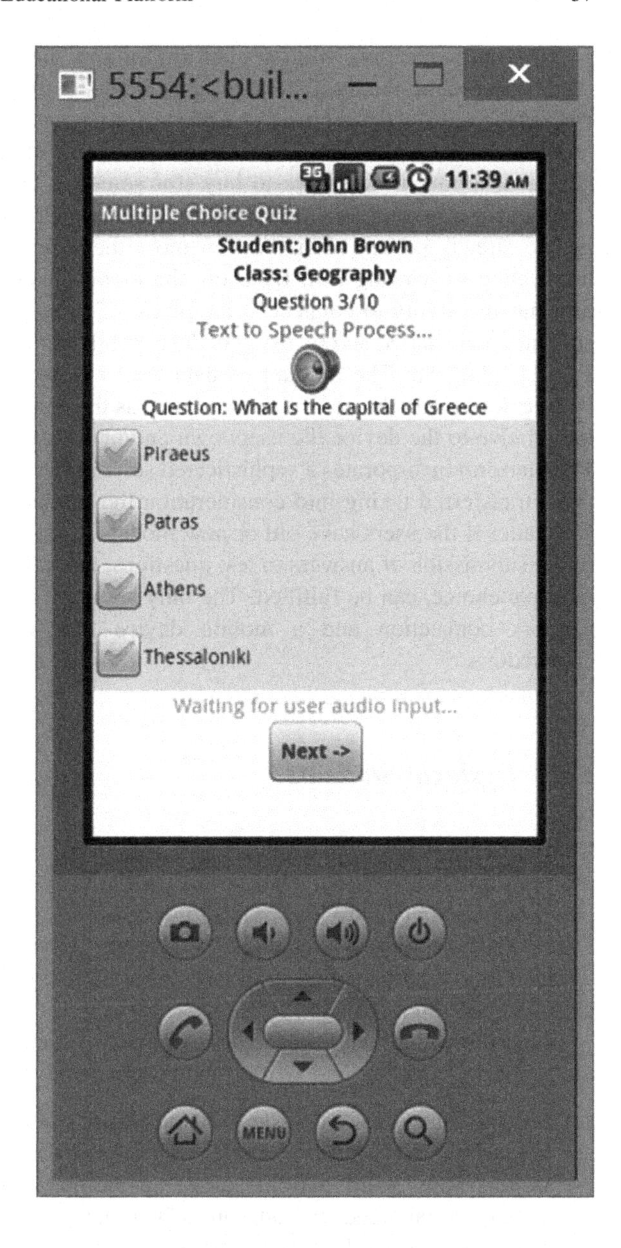

Fig. 5.7 Text-to-speech and speech recognition in the educational environment

oral interaction with the mobile application. Readers may compare Figs. 5.7 and 5.1 where the interaction is accomplished exclusively by keyboard/screen interaction. In figure K, the whole multiple choice question is spoken out loud, through a text-to-speech incorporated engine. Correspondingly, the user's input is also transmitted orally, through the audio modality, by using an incorporated speech recognizer.

The communication with the platform through the use of mobile devices could prove to be quite handy and useful for doctors as well as for their patients. For example, even though the user interface of a personal computer is in many ways (audio-visual, speed) beneficial in comparison with an interface of a mobile device, it is rather expensive to buy (for someone who doesn't have it). On the other hand, a mobile phone is considerably cheaper and is something that many people already possess. This is even more the case in countries where telecommunication is growing fast. By using the mobile facilities of the platform, every user has the ability to connect to the platform's main server that has the medical software preinstalled and by using his/her mobile device is able to interact with the medical platform. The amount of data that will be transferred to and from the mobile device and the main server, as well as the quality of the graphical interface are relative to the device the users own and/or the network they are connected to. The platform incorporates a sophisticated software mechanism that adjusts the data to be transferred taking into consideration the limitations of each mobile device. No matter if the users have old or new models of mobile devices, at least the part of the submission of answers to test questions, either by entering simple text or by multiple choice, can be fulfilled. The only assumption to achieve this is an active network connection and a mobile device that supports network (wireless) connections.

5.2.3 Dyslexic Students

The term "dyslexia", in a short form, means having difficulty with words and refers to the difficulty that a person has while reading, writing or spelling words that are either dictated to him or has to write on his own or copy them. Dyslexia is a brain malfunction that affects the way that the individual learns, which appears as a difficulty in reading and writing but also a difficulty in obtaining knowledge and skills in other areas of learning as well.

A more specified goal for the development of such targeted software would be the support of reading and writing with the general use of the hearing sense:

(1) Support during reading: The dyslexic person has the facility of "hearing" his/her written material with the help of a synthetic voice through speech engines. In this way, word and meaning misinterpretations from the reading can actually be avoided. In addition, it is easier for the majority of dyslexic persons to hear something rather than to try reading it, since besides everything else dyslexic people have been observed to have a weakness in the ability to focus while reading.
(2) Support during the written expression and especially during tests: While using the computer's keyboard, dyslexic students are observed to muddle up or invert some letters and/or numbers. Help is provided in such cases at many levels.

In order to recognize a mistake, the platform first compares the users' input with data from a dictionary of the language that the user uses. In this way, many orthographical mistakes may be located which are not necessarily due to dyslexia. The platform can then suggest the replacement of the wrong word with word(s) that look alike and are orthographically correct.

Aiming mainly at recognizing input errors from dyslexic users, the system incorporates a special facility of checking/comparing alphanumeric values, which is based on sound comparison. As it is already stated, dyslexic people may write something wrong and when they read it they cannot easily find what their mistake is. However, when they hear something that is read out loud, it is much easier for them to track changes or misconceptions from the audio channel of interaction. While using the platform input data is bi-modal, since users can interact with the system either orally (through the mobile microphone), or through the mobile keyboard and touch-screen. More specifically, while taking tests there is a special string comparison mechanism that takes place that compares the input data with the correct answer (stored in the systems database) using a sound comparison approach. Both the user's answer and the correct answer are transformed into phonemes using certain rules and then these phonemes are being compared. This method is opposed to the "classical" string comparison approach. The idea that lies behind this specific "check" is that in many situations wrong words are written in a way that sound the same as the correct ones. In addition, there are many cases where dyslexic users have been observed to have a tendency to exchange consonants. For the majority of such cases, the use of a wrong consonant does not change the sound and the vocalism of the word so much as it would if the wrong vowel was used. The approach we have used for error diagnosis within the educational application is illustrated in Fig. 5.8. As a first step the system collects the input data in text form (either directly from the mobile keyboard or after voice recognition from the mobile microphone) and then the text string is transformed into sound phonemes. The same procedure is followed for the data that is stored in the systems database (correct answers to tests) and finally the emerging phonemes are compared indicating whether they match or not.

In each case the platform provides the quite simple yet important potential of reading through the platform with the use of a speech engine for the text that the user types in real time. This means that the dyslexic user listens to what he/she types while typing through the keyboard on his/her personal computer. Thus, dyslexic students are helped by not being confused with visually similar words (and also semantically similar words) by hearing these words so that they may track their own mistakes more easily and more quickly. For example they could write or read "Pavlos" instead of "Paul" as "similar" words, because firstly they are names, secondly because visually they have a similar length and finally because both begin with the letter "P". However, if a speech engine reads out loud these two words to a dyslexic, the difference is much more easily revealed.

Furthermore, the platform offers the facility of the replacement of the keyboard as an input device by a microphone with the use of voice recognition. The user has the ability not only to input vocally but also to operate the program vocally. More

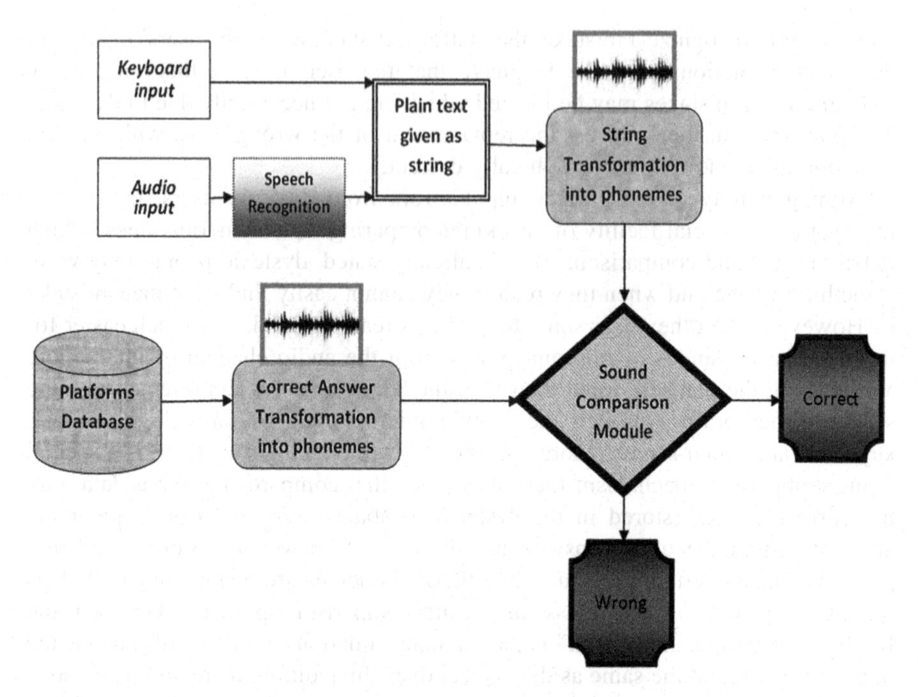

Fig. 5.8 Sound comparison mechanism for dyslexic mistakes checking

specifically, the platform teaches itself based on the user's voice and as a result the success ratio for correct input increases. Data input with the use of a user's own voice is something that can accommodate everyone, let alone dyslexics. All of the above facilities are based on incorporated theories and technologies. The nature of dyslexia necessitates the creation of a knowledge database, which would draw from and incorporate data based on dyslexia. Moreover, it could incorporate the various grammatical-rules and syntactic-rules which we have assigned to the specific dyslexic and could also write data to their personal knowledge database. For example, although we cannot specify as an exploitable by the platform rule that dyslexics always confuse or mess-up opposite concepts, we can however "instruct" the application with the information that a large percentage of dyslexics confuse "right" and "left" or "black" and "white".

Likewise, the platform can create "dyslexic user models" so that better results may be achieved for each individual user. The interaction with the platform is more individualized and adaptive to each dyslexic case. With the creation of dyslexic user models, the platform will have the ability to track the particular user's weaknesses and offer the specific help required for this user's needs.

Supervision of dyslexics during their interaction with the platform can also be achieved by doctors with the use of wireless communication via mobile phones. It is necessary for a professional to follow the progress of the dyslexics while they are interacting with the platform in order to evaluate the results of the interaction (benefits, problems, etc.) and also to be in a position to adapt the platform's parameters based on the users needs. Both instructors and doctors do not need to purchase a computer, desktop or laptop, but rather use their mobile device which they may already have. Additionally, the instructor is not required to have specific computer knowledge since the educational GUI (graphical user interface) is quite helpful and user-friendly. Access and user identification can be achieved at any moment and place and as many times required as one of the main advantages of mobile technology. Finally, with the use of this technology, the dyslexics are introduced to communication among themselves and the instructor(s) who can oversee them at any given moment from any given place.

5.3 Mobile Coordination of People Who Support Students with Special Needs

Students with special needs need help that should be provided not only from their family, but also from their school, specific organizations and foundations and the community indiscriminately. Comprehension and support for these students improve the more knowledge we have in each particular situation and are critically associated with the quality of services we can afford to give. These services are presented, analyzed and reviewed in (Chung et al. 1999), where we can clearly see the fact that deficient support of people with special needs has to do, to a large degree, with the missing, qualitative and quantitative, of the corresponding mediums and services from the community. Moreover, professional on-site help is particularly expensive and in most cases, either the government cannot provide it, or the patient's family cannot afford it.

Since provision of help from a professional or doctor is expensive and taking into account the fact that people with special needs need continuous observation, the software we are proposing is developed mainly to complement the work of a doctor or a consultant and also to provide important and useful information to the parents.

All users, students with special needs, therapists, tutors, parents and co-students can use the platform to cooperate and to interact with each other. All are not only able to have easy access to the mobile platform's databases, but they can also communicate with each other. Some usual cases of communication are the following:

1. A therapist sends comments to a tutor about a student who possibly suffers from dyslexia and needs special care. Specific tests and theory adapted to the student's problem.
2. A tutor informs a therapist that a specific student seems to have "unusual" problems in Greek grammar.
3. A tutor sends a message to a student's parent to inform him/her about their child's performance.

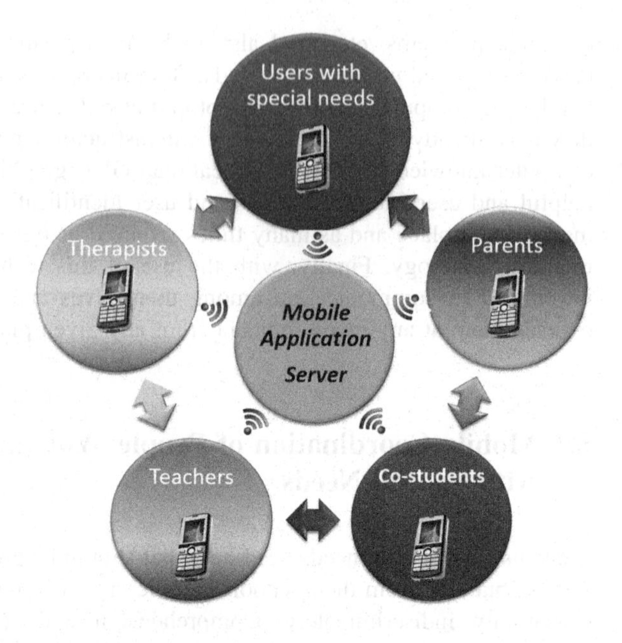

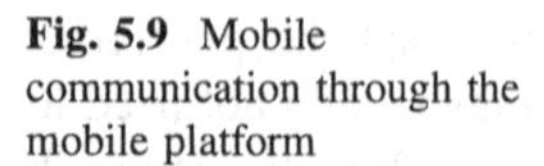
Fig. 5.9 Mobile communication through the mobile platform

4. A tutor sends a message directly to a student as advice to try to be more concentrated while solving multiple choice exercises.
5. A student sends messages to a tutor about a misconception she/he has while using the mobile educational platform.
6. Fellow students send messages to each other in terms of collaboration.

The communication between them can be realized in many ways as it is illustrated in Fig. 5.9. By using their mobile phone (and of course connecting to the platform's mobile pages) therapists can send short messages via the short message service (SMS), either to the students or to their tutors and parents (If they also have mobile phones). The email service can be used as well in situations where something more than simple "short messages" is needed. Alternatively therapists can write the messages to the platform's data-base. In this case, therapists have to declare the name of the receiver (whether she/he is a tutor, a parent or a student) and the platform will use its audio-visual interface to inform him/her as soon as she/he opens the platform.

In the first case (e-mail or SMS message) the message is written to the platform's database and then is sent to an internet service that provides SMS and e-mail sending. Therapists are also able to send an e-mail directly through their mobile phones but it may be preferable for them to use the platform to do that. The main reason for this is the fact that mobile networks are considerably slow and cost much and thus they may not be very convenient for the platform's users. Thus the platform is expected to send an e-mail to the internet service. The body of this e-mail will be the body of the short message and the "subject" of the e-mail will specify the receiver by his/her mobile telephone number. Messages (if sent from the tutor to the student for example) can contain information about which test the student should take next or anything that the tutor might think that the student should pay attention at. Messages can also be sent to parents (from therapists or tutors) who may find an easy way to stay informed about their student's progress.

Additionally, even greater functionality is given to the owners of the technologically latest mobile phones (Smartphones). They may use the quite recent but rapidly spreading technology of Bluetooth or Wi-Fi wireless communication. This means that they may alternatively use the above technologies in order to send and/or receive data to and from the main server of the educational platform through their mobile device. Of course there is a limitation in the distance between their device and the server (with Bluetooth approximately 10–20 m), but considering the actually inexpensive way of communication, most users may find it rather useful.

The discrimination between the roles of the users is done in the main server and for each different user a personal profile is created and stored in the data base. In the data base, other personal information, such as e-mail accounts and mobile phone numbers are stored so that users won't need to recall them while trying to communicate with each other. Only a private key such as the user's name is needed in most cases. In order to accomplish this, the user name and password is required when logging in the mobile application to gain access to the mobile interface. On the first level of authentication the platform determines whether there is a valid username-password or not. On the second level of authentication users are distinguished in users of the fore mentioned basic categories, namely students, tutors, parents and therapists. A note on the security and privacy aspects of the mobile platform, the platform supports the use of the Secure Sockets Layer (SSL) in order to encrypt data during important data transactions.

5.4 Conclusions

This chapter has presented a mobile educational platform that supports the education of students with special needs. The platform involves facilities that allow its use by students with disabilities and also records their progress and common weaknesses in an effort to become more personalized for them. Moreover, one important feature of the platform is that it provides support, communication and coordination of other people involved through the platform, to reinforce efficient supervision. The platform's ultimate aim is the inclusion of the students with special needs in the social context of learning.

References

Avramidis E, Bayliss P, Burden R (2000) Student tutor's attitudes towards the inclusion of students with special educational needs in the ordinary school. Teach Tutor Educ 16:277–293

Barbeau SJ, Winters PL, Georggi NL, Labrador MA, Perez R (2010) Travel assistance device: utilising global positioning system-enabled mobile phones to aid transit riders with special needs. IET Intel Transp Syst 4(1):12–23

Barrett JC (2000) A school-based care management service for students with special needs. Fam Community Health 23(2):36–42

Bertini E, Kimani S (2003) Mobile devices: opportunities for users with special needs. In: Lecture notes in computer science (including subseries Lecture notes in artificial intelligence and Lecture notes in bioinformatics), vol 2795, Springer, Berlin, pp 486–491

Brooks C, Miller LC, Dane J, Perkins D, Bullock L, Libbus MK, Johnson P, Van Stone J (2002) Program evaluation of mobile dental services for children with special health care needs. Spec Care Dent Off Publ Am Assoc Hosp Dent Acad Dent Handicap Am Soc Geriatr Dent 22(4):156–160

Chung MC, Vostanis P, Cumella S, Doran J, Winchester C, Wun WL (1999) Students with special needs: use of health services, behaviour and ethnicity. Stud Youth Serv Rev 21(5):413–426

Hasselbring TS, Glaser CHW (2000) Use of computer technology to help students with special needs. Future Stud 10(2):102–122

Kracker MJ (2000) Classroom discourse: teaching, learning, and learning disabilities. Teach Tutor Educ 16:295–313

Monibi M, Hayes GR (2008) Mocotos: mobile communications tools for children with special needs. In: Proceedings of the 7th international conference on interaction design and children, IDC 2008, pp 121–124

Pearson V, Lo E, Chui E, Wong D (2003) A heart to learn and care? Tutors' responses toward special needs students in mainstream schools in Hong Kong. Disabil Soc 18(4):489–508

Sullivan HT, Häkkinen MT, DeBlois K (2010) Communicating critical information using mobile phones to populations with special needs. Int J Emerg Manage 7(1):6–16

Chapter 6
Mobile Versus Desktop Educational Applications

Abstract Software that is meant to help an educational process can be considered successful only if it is accepted and approved by the interacting targeted participants. For educational software, there are two basic groups of users who use it, namely instructors and students. As a result, in this chapter, we try to find out how helpful our mobile learning software might be for human instructors and also how educationally beneficial it is to their students. On a second level, our evaluation is targeted to measure the effectiveness of a mobile approach to learning compared with the "traditional" computer based, e-learning process.

6.1 Introduction

Educational applications constitute a special category of software that needs to model and assist many aspects of the cognitive processes of humans whether these are learners or instructors. Formative evaluation is one of the most critical steps in the development of learning materials because it helps the designer improve the cost-effectiveness of the software and this increases the likelihood that the final product will achieve its stated goals (Chou 1999). In the literature there are evaluation methods that are completely specialized to educational software. One such evaluation framework outlines three basic dimensions that need to be evaluated: (i) context; (ii) interactions; and (iii) attitudes and outcomes (Jones et al. 1999). The context determines the reason why the educational software is adopted in the first place, i.e. the underlying rationale for its development and use; different rationales require different evaluation approaches. Students' interactions with the software reveal information about the students' learning processes. The "outcomes" stage examines information from a variety of sources, such as pre and post-achievement tests, interviews and questionnaires with students and tutors. This framework has been used for the evaluation of our mobile learning software platform MAT, that was presented in Chap. 4.

E. Alepis and M. Virvou, *Object-Oriented User Interfaces for Personalized Mobile Learning*, Intelligent Systems Reference Library 64,
DOI: 10.1007/978-3-642-53851-3_6, © Springer-Verlag Berlin Heidelberg 2014

The underlying rationale of MAT involves offering more convenience with respect to time, place and kind of device to its users (instructors and learners), therefore the context of the evaluation required an emphasis on the mobile aspect of the application. Then, students' interactions with the software were evaluated with respect to the students' learning processes while they used mobile devices. Finally, the "outcomes" stage involved pre and post-achievement tests before and after the use of a mobile device. In addition, it involved many interviews of students and instructors, which focused mainly on evaluating the use of mobile devices. In view of these, the evaluation of Mobile Authoring Tool (MAT) involved both instructors and students and was conducted in two different phases.

6.2 Settings of the Evaluation

At the first phase, the authoring procedure was evaluated by instructors, who were interviewed after they had developed an ITS. The second phase concerned the evaluation of the resulting educational applications and involved both instructors and students. The instructors of the second phase were exactly the same as in the first phase, so that they could have a complete experience with MAT, both for the creation of an ITS and the management of their course.

At the first and second phase, 25 instructors participated in the evaluation. Ten of them were secondary school instructors and were asked to prepare lessons and tests in geography, history, biology, physics, chemistry and English respectively depending on their expertise domain. The rest of the instructors were University instructors. Seven were Medical Science instructors while the other eight were Computer Science instructors. All of the instructors who participated in the experiment were familiar with the use of computers and to a smaller extend with mobile phones. In addition, they had been trained for the use of MAT before the experiment.

When interviewed, all of the instructors confirmed that MAT had a user-friendly interface and that the mobile facilities were either useful or very useful. More specifically, 22 of them stated that they found the mobile facilities of MAT either useful or very useful both for the creation and the maintenance of their courses whereas three of them said that they had not used the mobile features at all during the creation of the course but they found them useful during the maintenance of the course. As expected, all instructors who found useful the mobile features of the application for both phases, made clear that they had used the mobile facilities in a complementary way with a desktop computer, since the authoring process involves inserting a lot of data. Thus it would have been difficult for anyone to develop the whole course using a mobile phone.

It must be noted that among the most enthusiastic instructors about the mobile features were the University instructors who were Computer Science and Medical instructors. To some extent, it was expected that Computer Science instructors would probably like MAT more than the other experts due to their familiarity with technology. Indeed they rated the usefulness of the mobile features very highly. However, the application was even more appreciated by Medical Science Instructors. This was probably due to the fact that Medical Science Instructors usually depended a lot on their mobile phones due to the heavy responsibilities that they had and the fact that they had to be in many different places during the day.

The second phase of the evaluation study involved in total 50 students, two students selected from each of the respective classes of the 25 instructors who participated in the evaluation. The underlying rationale of mobile ITSs lies on the hypothesis that these applications are more convenient and flexible to use while they retain the educational quality. At a first glance, the validity of this hypothesis might look obvious. However, there may be students who are not familiar with educational software in general and thus might not like the particular applications. On the other hand, there may be students, who are very familiar with computers and mobile phones and are very happy to use them for educational purposes. Hence, one important aspect of the evaluation was to find out whether students were indeed helped by the mobile environment. Another very important aspect was to find out whether students had gained educational benefits from the ITSs.

Students were asked to use the resulting mobile ITS as part of their homework for their courses. After the courses were completed, the students were interviewed. Most of the students found the whole educational software application useful or very useful. Even greater was the appreciation of the instructor-students mobile collaboration through messages during the courses. The students who were more enthusiastic about these facilities were either the students who attended many lectures during the day and thus they had a very "mobile" timetable, or students who liked to use short message texting in their everyday lives for all types of their communication with other people. In particular, for this category of students, the facility of using mobile devices rendered the whole educational software application more attractive and engaging simply because they liked the mobile phone as a medium.

6.3 Evaluation Study for Students

As mentioned in the previous section, the evaluation study was conducted among 50 first-year students. These students were selected randomly from their classes. After the students were selected they were given a short interview concerning their computer knowledge and skills. The evaluation study of MAT concerned two levels:

1. Usefulness level. In this level, the usefulness of the desktop and the mobile facilities of the m-learning systems were evaluated.
2. User friendliness level. In this level, the user friendliness of the desktop and the mobile facilities of the m-learning systems were evaluated.

A sample of the questions that were asked to the students is the following:
Concerning the first level of the evaluation:

1. How do you rate the usefulness of the educational application?
2. How do you rate the usefulness of the mobile communication facilities?
3. Do you prefer using your mobile phone, rather than a PC for the interaction with the application? If yes/no what did/didn't you "like" most?

Concerning the second level of the evaluation:

1. Do you think that the "mobile" features prevented you from understanding the educational process better?
2. Did you consider the application attractive? If yes, what did you like about the application?

We observed that although most of the asked students preferred to use their mobile phone for the interaction, most of them where users with insufficient computer knowledge. On the other hand most of the users that had sufficient computer knowledge preferred them for the interaction.

The study also revealed that the degree of user friendliness with the software depended also on the available mobile device. However most of the "not satisfied" claimed that improvements of the software would probably balance the hardware deficiencies.

6.3.1 Evaluation Results

The percentages of the student's answers to the main questions of the evaluation study are illustrated in Tables 6.1, 6.2, 6.3 and 6.4.

As a result of this evaluation study, students have appreciated both the desktop and mobile features of the e-learning system for different reasons. The desktop facilities were considered very user-friendly by students who had previous computing experience. However, one very important finding came up from students who were not familiar with computers. Mobile facilities were preferred by these students. Most of the students, who do not have much experience in using computers, own a mobile phone and therefore know how to use it. These reasons make the mobile interaction "more attractive" and "accepted" by the majority of medical students. As to the user-friendliness of the mobile facilities, it was made clear that its level depended on the available mobile device. However most of the students suggested that there should be improvements of the software in order to improve the mobile communication for not user-friendly devices.

Table 6.1 Percentages of answers concerning the usefulness of the educational application

Usefulness of the educational application		
Very useful	Useful	Not useful
20 %	45 %	35 %

Table 6.2 Percentages of answers concerning the usefulness of the mobile interaction

Usefulness of the mobile communication		
Very useful	Useful	Not useful
30 %	45 %	25 %

Table 6.3 Percentages of answers concerning the preferable device for the interaction

Preferable device for the interaction				
Mobile phone		Personal computer		Both
48 %		30 %		22 %
Users with computer knowledge	Users without computer knowledge	Users with computer knowledge	Users without computer knowledge	
30 %	70 %	80 %	20 %	

Table 6.4 Percentages of answers concerning the easiness in navigation through the mobile pages

Easiness in navigation through the mobile pages		
Quite easy	Not easy	Device depended
26 %	20 %	54 %
	Do you think that the appropriate software might improve the situation?	
	Yes	No
	85 %	15 %

6.4 Evaluation Study for Instructors

To find out to what extent the usefulness and the usability of the mobile facilities were, we extended our evaluation study to instructors. In general, the majority of instructors found the mobile facilities useful and only the minority of them had not used them at all. The present evaluation study focused on the usefulness and the usability of the mobile features as compared to the standard desktop devices. Therefore an important question was whether instructors were helped by the mobile facilities and whether they found them easy to use in comparison with the desktop web-based facilities of the application.

The evaluation study that we describe in this section involved 25 instructors of various domains. The selection of the instructors was made in such a way so that almost one half (11) of the instructors were adequately experienced with the use of computers and the other (14) of the instructors had no significant experience with the use of computers. The inexperienced section of instructors had also been divided into two groups: One group of instructors participated in a short seminar concerning new technologies and the use of mobile devices and another part did not attend the seminar. The reason for this kind of grouping of instructors was to find out whether mobile facilities can provide support to users irrespective of their computer knowledge and their possibility of accessing desktop computer equipment.

During the evaluation study all aforementioned 25 instructors were given access to desktop computing facilities and mobile phones and were asked to author and manage their courses using both desktop and mobile devices. Afterwards, they were asked questions about their experiences.

6.5 Usefulness of the System's Features

To find out how useful the mobile facilities were all the instructors were asked questions concerning the usefulness of the mobile features of MAT. A sample of the questions that the 25 instructors were asked is the following:

1. How do you rate the usefulness of the authoring-instructing application?
2. How do you rate the usefulness of the mobile communication facilities?
3. Do you prefer using your mobile phone, rather than a PC for the interaction with the application? If yes, in which cases do you think that a "mobile interaction" is preferred?
4. Do you think that it is possible to "administer" a whole lesson from a handheld device?
5. Do you think that the alternative use of your mobile phone in the whole educational process could save you precious time?
6. Do you think that the "need" of incorporating mobile features in educational applications will increase in the future? If yes, why?

A summary of some important results is illustrated in Table 6.5. The majority of instructors rated the whole authoring-instructing process as useful or very useful. We must note that instructors with heavily loaded schedules found the idea of an alternative "mobile" interaction quite attractive. That results from the fact that the authoring-instructing process could take place from anywhere at any time. Another very important result is that the mobile features are mostly preferred by instructors with insufficient computer knowledge. Most of the instructors, who do not have much experience in using computers, own a mobile phone and therefore know how to use it. Of course using a mobile phone is easier than a computer, even in modern user-friendly interfaces. On the other hand, most of the instructors that where familiar with the use of computers preferred computers for the interaction. The reason for this was that computers provide a whole variety of facilities

Table 6.5 A summary of parts of the results

Preferable device for the administration of a whole lesson				
Smartphone		Desktop computer		Both
16 %		24 %		60 %
Preferable device for the interaction				
Smartphone		Desktop computer		Both
40 %		20 %		40 %
Users with computer knowledge	Users without computer knowledge	Users with computer knowledge	Users without computer knowledge	
20 %	80 %	76 %	24 %	

Table 6.6 Usability in navigation through mobile forms

Easiness in navigation during mobile interaction		
Quite easy	Not easy	Device depended
32 %	16 %	52 %

such as large keyboards, mouse and user-friendly menus. Thus, instructors who already knew how to use computers preferred the computers for their standard interaction with the application. Finally, almost all of them believed that there will be a significant increase in the "need" for mobile potential in the near future.

6.6 Usability of the System's Features

This part of the evaluation aimed at finding out how usable the mobile facilities were both on their own stand and in comparison with standard desktop computers. A sample of questions asked is the following:

1. Did you find it easy "navigating" through the mobile forms?
2. Please rate the easiness of use of the mobile facilities and then the computer-based facilities.

In Table 6.6 we can see that one third of the instructors found it easy to use the mobile capabilities. Both mobile and computer-based facilities were quite easy in use. Of course mobile facilities differed from the computer-based facilities, as they were less sophisticated but not so easy to use. It was made clear that the level of user friendliness depended on the available mobile device and of course on the experience of the instructor in using such applications. However most of the less satisfied users claimed that improvements of the software would probably balance the hardware deficiencies.

References

Chou C (1999) Developing CLUE: a formative evaluation system for computer network learning courseware, J Interact Learn Res 10(2):179–193

Jones A, Scanlon E, Tosunoglu C, Morris E, Ross S, Butcher P, Greenberg J (1999) Contexts for evaluating educational software. Interact Comput 11(5):499–516

Chapter 7
Multiple Modalities in Mobile Interfaces

Abstract In human–computer interaction, a modality refers to a path of communication between the human and the computer, such as audition and vision. Computers, embedded computer systems and especially modern mobile phones (also known as smartphones) have a growing "tendency" to add more paths of communication between them and the outer world. In this chapter the authors give a thorough overview of the multiple modalities of interaction between smartphones and humans that can be found in recent mobile devices and features that may be expected to make their entrance in the mobile technology in the near future.

7.1 Introduction

Mobile interaction modalities include specific sensors that may also be considered as modalities of interaction since they provide these devices with the ability to interact with their environment more sophisticatedly than the usual interaction. Some common modalities include the mobile keyboard, the mobile microphone and speaker and the mobile screen. These modalities are used mostly by humans for their bi-directional interaction, input and output of data, with their mobile devices.

Other quite common modalities of interaction include cameras that many mobile devices have, external Bluetooth devices that can be connected to most of the available mobile phones and touch or multi-touch screens that can be used also as modalities of exchanging data between users and their mobile devices. However, there is also a quite a big number of less common and even rarer modalities and sensors that are incorporated into high tech mobile devices and smartphones that enable them to collect even more data from their environment for the needs of their potential users and/or services. For example, a GPS (Global Positioning System) sensor can give very important and useful information about the geographic location of a device and its corresponding user and as a result this

E. Alepis and M. Virvou, *Object-Oriented User Interfaces for Personalized Mobile Learning*, Intelligent Systems Reference Library 64,
DOI: 10.1007/978-3-642-53851-3_7, © Springer-Verlag Berlin Heidelberg 2014

Fig. 7.1 Smartphone sensors and applications

information can be used by a variety of geolocation applications that are quite widespread nowadays. A less common example of a sensor is that of the accelerometer sensor that is used to measure the acceleration and also the orientation of a mobile device. In this chapter we give an overview of these modalities of interaction as well as short descriptions of potential modalities than can or will be used by the mobile technology in the near future. It is of great importance for all software technologists to be aware of these technological aspects so that they can design applications that may maximize their utilization with their environment. Figure 7.1 illustrates a smartphone that uses its sensors to interact in multiple ways and dimensions with its users and its environment through the use of mobile applications.

7.2 Recent Works in Smartphone Sensors

The authors of (Zhang et al. 2013) present a prototype on an Android smartphone that can sample the related sensors during the user's movement and collect the sensor data for further processing on personal computers. The work of (Ahmetovic 2013) analyzes the challenges in unassisted orientation and way-finding, especially in unexplored and potentially dangerous environments for visually impaired users. Prudêncio et al. (2013) propose a novel set of features for distinguishing five physical activities using only sensors embedded in the smartphone by introducing features that are normalized using the orientation sensor. The authors of (Mehta et al. 2012) report on the development of a new, versatile, and cost-effective

Fig. 7.2 Classic mobile keyboard

clinical tool for mobile voice monitoring that acquires the high-bandwidth signal from an accelerometer sensor placed on the neck skin above the collarbone. A mobile learning system that incorporates smartphone sensors is found in (Tsai and Peng 2011). This system's authors propose a portable and convenient learning assisted system by using Android Smartphone with wireless sensors. The resulting system senses and collects the data of learning behaviors with a smartphone as the processing unit.

7.3 Common Modalities of Interaction in Smartphones

In this section we examine the most common modalities of interaction in smartphones which can also be found in most mobile devices around the word.

7.3.1 Mobile Keyboard

A mobile keyboard is a portable keyboard that is designed to be used with wireless devices, such as mobile phones, personal digital assistants (PDAs) and smartphones. A mobile keyboard has reduced size and also a smaller number of buttons in comparison with a standard desktop computer keyboard. There are at least three types of mobile keyboard types. The classical mobile keyboard that is illustrated in Fig. 7.2, a QWERTY style mobile keyboard, illustrated in Fig. 7.3 and a touch screen mobile keyboard that can also change its appearance, illustrated in Fig. 7.4.

Fig. 7.3 A horizontal QWERTY mobile keyboard

Fig. 7.4 On screen touch keyboard

7.3.2 Mobile Microphone

A mobile microphone is actually an acoustic-to-electric sensor that is responsible to convert sound from a mobile device's environment into an electrical signal that can be processed, stored or transmitted through a mobile device. A typical mobile

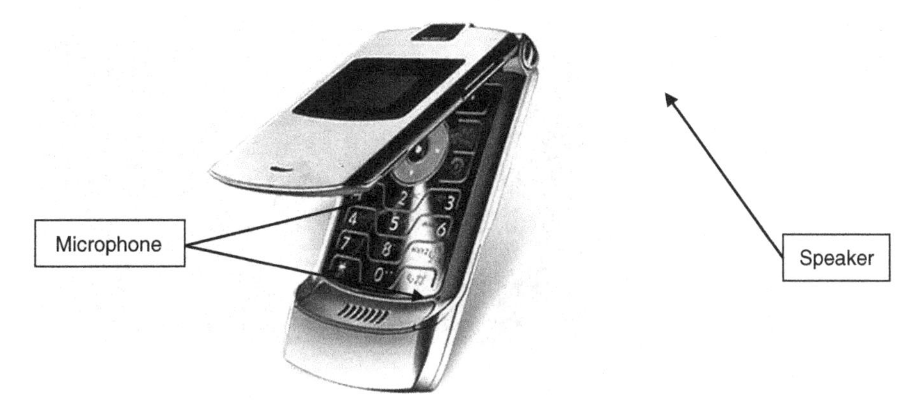

Fig. 7.5 Mobile microphone and speaker

Fig. 7.6 Mobile handsfree and mobile Bluetooth device

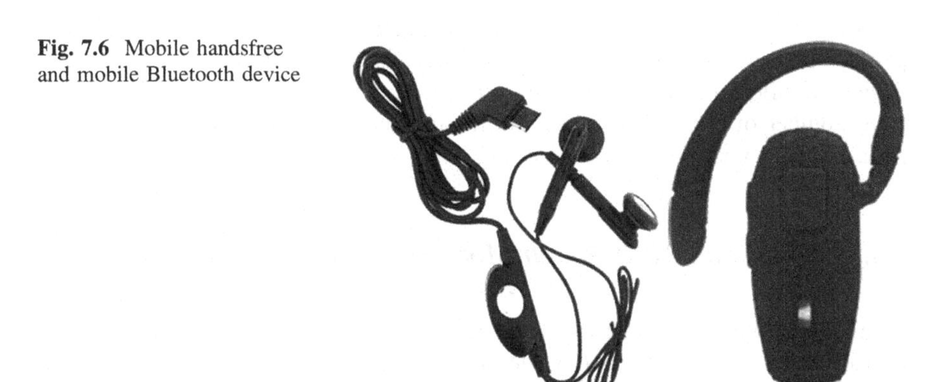

microphone is illustrated in Fig. 7.5, while Fig. 7.6 illustrates mobile microphones that operate through the use of hands-free cabled and Bluetooth devices.

7.3.3 Mobile Speaker

A mobile speaker is the functional opposite of the mobile microphone. A mobile speaker is the device that enables the acoustic communication of a mobile device and its environment by transforming the mobile device's output into sound waves. It is mostly used as a basic interaction device between a phone and a user when making phone calls, but it is also widely used as a multimedia interaction modality through software mobile applications. A typical mobile speaker is illustrated in Fig. 7.5, while Fig. 7.6 illustrates mobile speakers that operate through the use of hands-free cabled and Bluetooth devices.

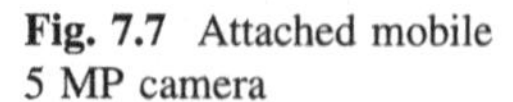

Fig. 7.7 Attached mobile 5 MP camera

7.3.4 Mobile Camera

A mobile camera is a device that gives a mobile phone the ability to capture still photographs and usually also video in a variety of resolutions depending on each device. Mobile cameras are currently used as input modalities in phone calls, in a large number of research projects and in commercial applications. A 5.0 Megapixel mobile camera is illustrated in Fig. 7.7.

7.3.5 Touch and Multi-touch Mobile Displays

Multi-touch mobile displays are the evolution of simple touch displays, which provide the sensing surface (mobile screen) with the ability to recognize the presence of two or more points of contact with the surface of the display, while a single touch or touch display recognizes the presence of a single point of contact. Multi-touch mobile displays have many useful applications such as easily zooming in and out and also rotating on-screen objects. A growing number of multi-touch displays also support measurements for user finger pressure. A mobile phone with a touch display is illustrated in Fig. 7.8.

7.4 Sensors Found in Modern Smartphone Devices

7.4.1 Wi-Fi and Bluetooth (Can be Used Both for Communication and for Sensing Wireless Signals)

Bluetooth and Wi-Fi have many applications: setting up networks, printing, or transferring images and other files. Wi-Fi is intended as a replacement for cabling for general local area network access, while Bluetooth is used for exchanging data over short distances from fixed and mobile devices (usually from 0–20 m), creating personal area networks with high levels of security.

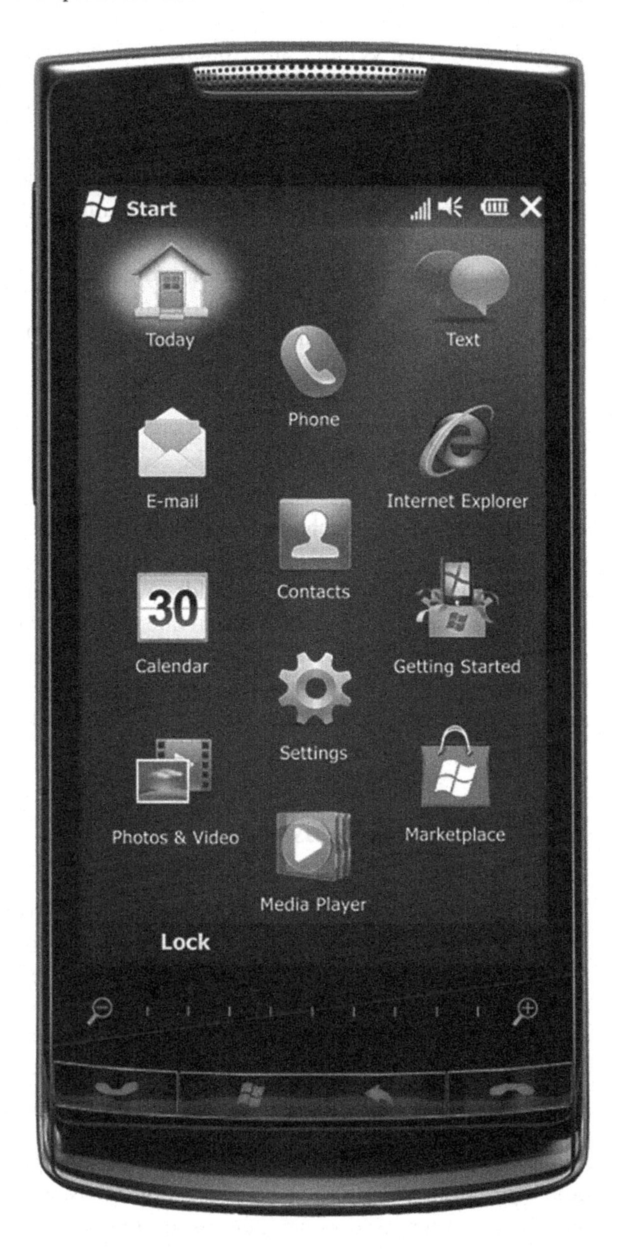

Fig. 7.8 Standard mobile multi-touch screen

7.4.2 GPS

Current regulations (2013) are encouraging mobile phone tracking, for this reason the majority of GPS receivers are built into mobile telephones, with varying degrees of coverage and user accessibility. Navigation software is available for most Twenty First century smartphones and also for some Java-enabled phones that allow them to use an internal or external GPS receiver. Many modern mobile phones which incorporate GPS work by assisted GPS (A-GPS) only, which means that can only function in conjunction with their carrier's cell towers. Other mobile devices can provide navigation or location capabilities with satellite GPS signals as dedicated portable GPS receivers. Finally, other models also exist which have a hybrid positioning system that can use other signals when GPS signals are inadequate (such as magnetic field sensor and accelerometer sensor). GPS allow both GPS navigation as well as localization applications. A GPS software application operating in a mobile phone is illustrated in Fig. 7.9.

7.4.3 Proximity Sensor

A proximity sensor attached to a mobile device gives the device the ability to detect the presence of nearby objects without any physical contact, by using electromagnetic or electrostatic fields. As an example of this sensor's use, a proximity sensor can be used to deactivate the mobile phone's display and/or touch screen when the device is brought near a user's face (which is detected by the proximity sensor) during a phone call. This way we may save battery power and also prevent inadvertent inputs from users' face and ears.

7.4.4 Orientation Sensor

An orientation sensor senses a mobile device's vertical or horizontal orientation. There are two vertical positions that can be detected by an orientation sensor, namely the 90 degree angle and the 270 degree angle. There are also two horizontal positions that can be detected by an orientation sensor, namely the 0 degree angle and the 180 degree angle. Sensing a mobile device's orientation has several applications that include rotating clockwise or counterclockwise a web page while browsing and wanting to change the mobile display's orientation, viewing text files (adjusting to the longer or shorter dimension of the mobile device's screen, thus making the text much more legible) and taking photos through cameras with a user's preferable orientation by presenting landscape or portrait views. Accelerometers are often used as orientation sensors.

7.4.5 Magnetic Field Sensor

Magnetic field sensors (also known as smartphones' compasses) in mobile devices are used to measure the strength and/or direction of a magnetic field, produced

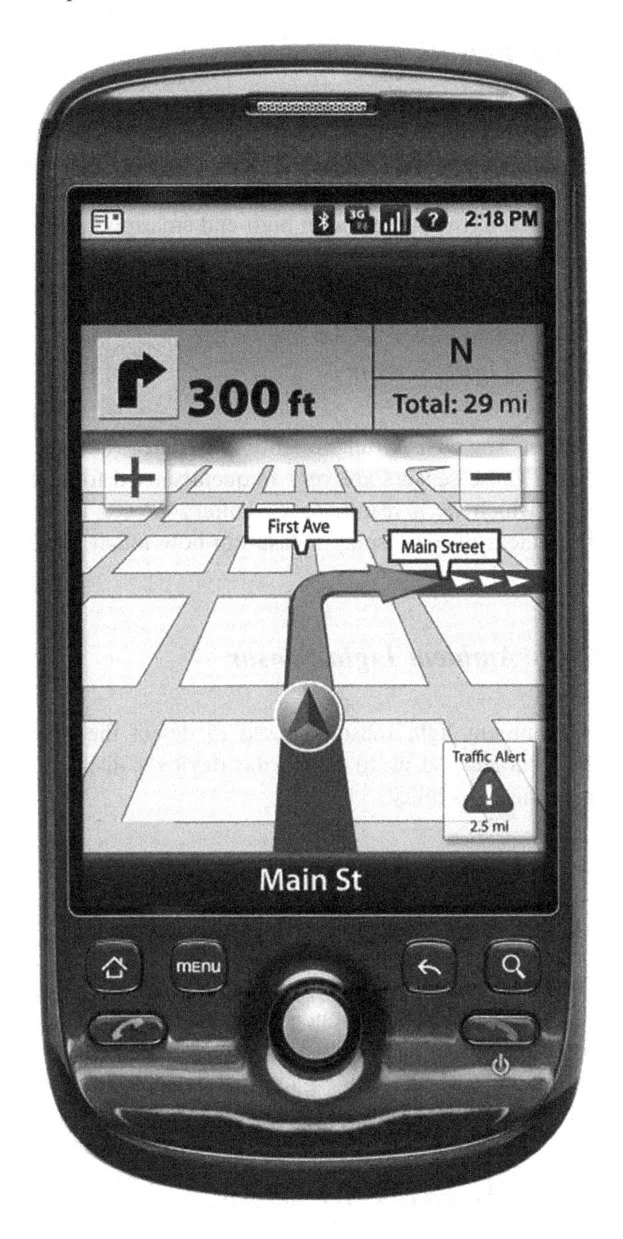

Fig. 7.9 Location services through GPS sensor

either by technical means or existing in nature. Magnetometers embedded in mobile devices may also be used in order to permit touch-less user-mobile device interaction.

7.4.6 Flashlight

A simple light or a more advanced flashlight are not genuine sensors but are very frequently used in combination with other mobile sensors. Their most well-known use is to cooperate with the smartphone's camera. A flashlight is an electric-powered light source and in high-end smartphones the light source is a LED (light emitting diode).

7.4.7 Light Sensor

This sensor's basic functionality is to detect and measure the valence of ambient light. These sensors are very frequently used to adjust the mobile display brightness which as a result, saves battery power in Smartphones by adjusting the brightness of the display measuring how much ambient light is present.

7.4.8 Ambient Light Sensor

An ambient light sensor is used to detect the brightness of a mobile device's environment, so as to adjust the device's display to bright or dim in order to maximize visibility.

7.4.9 Tilt Sensor

A tilt sensor embedded in a mobile device can be used to measure the tilting in often two axes of a reference plane in two axes. Accelerometer and tilt sensors are used in modern applications as common modalities for user-mobile phone interaction and also in mobile games as game controllers.

7.4.10 Accelerometer Sensor

An accelerometer is a sensor which can be found in mobile devices that measures the proper acceleration of the device. The accelerometer is also used in order to measure the orientation or vertical and horizontal positioning of the phone.

7.5 Less Common Sensors that Can be Found in Recent (2013) Smartphone Devices

7.5.1 Gravity Sensor

A gravity sensor is a device that attempts to measure the acceleration effect of Earth's gravity on the device. This measurement is typically derived by using the accelerometer, while other sensors such as the magnetometer and the gyroscope help to remove linear acceleration from the input data.

7.5.2 Gyroscope Sensor

As with the orientation sensor, the gyroscope is used in order to handle and maintain orientation. Gyroscopes are generally more advanced sensors than the orientation sensors. Gyroscope's main mechanism functions by the principles of angular momentum. These sensors are very frequently used in navigation systems and also in the fast growing field of gesture recognition. These sensors are proposed to be used in medicine, incorporated in the human body.

7.5.3 Pressure Sensor

A pressure sensor is used to measure finger pressure on a mobile device's touch screen, which means how hard a user is pressing a touch mobile display. Some devices have built in atmospheric pressure sensor in order to measure atmospheric pressure. However, these kinds of sensors are referred as barometer sensors.

7.5.4 Temperature Sensor

This sensor is our well-known thermometer which is attached inside the main body of a smartphone. It is used to measure environmental temperature. One important obstacle that has to be surpassed is the fact that each smartphone's internal hardware produces heat (variable to the device's use) and thus may influence the external temperature measurement.

7.5.5 Barometer Sensor

This sensor is already incorporated in high-tech smartphones and its purpose is to measure atmospheric pressure. Measuring pressure can be used to forecast short term changes in the weather and can be also used to estimate altitude.

7.5.6 Altimeter Sensor

A temperature sensor can be used to measure the external temperature of the environment of a mobile device. Temperature sensors also exist in other built in mobile components such as batteries in order to give useful information about the internal mobile system's temperature. Exceeding certain levels of temperature may result in hardware and software malfunctioning or even in hardware damage.

7.6 Future Sensors that Can be Embedded in Smartphone Devices

7.6.1 Perspiration Sensor

Perspiration sensors will be used in future smartphones in order to detect perspiration and could be used to monitor peoples' excitement level and even affective states.

7.6.2 User Body-Temperature Sensor

This sensor is almost identical to the thermometer. The only difference is that it is supposed to measure user's body temperature rather than environmental temperature.

7.6.3 Humidity Sensor (Also Known as Hygrometer)

A humidity sensor is used to measure environmental humidity. Along with the thermometer these two sensors give the smartphone the capability to sense its environment, thus become a weather station that can read weather signals.

7.6.4 User Blood Oxygen Level Sensor

This is a sensor that is proposed to function as a blood oxygen level measurement device. While it is proposed, we still do not have much information about its implementation.

7.6.5 Heart-Rate Sensor

This sensor is supposed to work as a heart-rate device that will measure the smartphone user's heart rate. We may assume that this sensor will be attached to the smartphone and will need user's skin contact in order to make accurate measurements.

7.6.6 Smell Sensors

This is also a proposed sensor to be incorporated to future smartphones. This sensor will be capable of sensing specific smells. Though such an incorporation may seem questionable, there are already many proposals for this sensor's use such as fire prevention through smoke detection.

7.7 Conclusions

Sensors can sense a large number of environmental changes and user input actions and make mobile devices respond automatically in order to improve user experience and also for device specific technical reasons such as improving mobile devices' battery life. There are cases where two or more modalities may combine their sensing or output capabilities in order to provide even more sophisticated sensing. For example a mobile camera attached on a human finger combined with a turned on mobile led light source may give enough information so that an application can measure a user's heart rate through changes in his/her skin color and blood. Even more intelligent applications can be used as human emotion detectors by combining tri-modal information from mobile cameras, mobile keyboards and mobile microphones. All these sensors, as modalities, that have been presented in this chapter have shown such a great potential and have found numerous applications over the very recent years that the technology of desktop personal computers and laptop computers has also expressed interest in incorporating them in future computer and operating systems.

References

Ahmetovic D (2013) Smartphone-assisted mobility in urban environments for visually impaired users through computer vision and sensor fusion. In: Proceedings—IEEE international conference on mobile data management, vol 2, article number 6569055, pp 15–18

Mehta DD, Zañartu M, Feng SW, Cheyne HAI, Hillman RE (2012) Mobile voice health monitoring using a wearable accelerometer sensor and a smartphone platform. IEEE Trans Biomed Eng 59(12 PART2):3090–3096 (article number 6257444)

Prudêncio J, Aguiar A, Lucani D (2013) Physical activity recognition from smartphone embedded sensors. Lecture notes in computer science (including subseries lecture notes in artificial intelligence and lecture notes in bioinformatics), vol 7887 LNCS, Springer, Berlin, pp 863–872

Tsai T-C, Peng C-T (2011) A smartphone assisted learning system with wireless sensors. Commun Comput Inf Sci (CCIS) 216(PART 3):557–561

Zhang L, Liu J, Jiang H, Guan Y (2013) Senstrack: energy-efficient location tracking with smartphone sensors. IEEE Sens J 13(10):3775–3784 (article number 6563204)

Chapter 8
Object Oriented Design for Multiple Modalities in Affective Interaction

Abstract The purpose of this chapter is to investigate how an object oriented (OO) architecture can be adapted to cope with multimodal emotion recognition applications with mobile interfaces. A large obstacle in this direction is the fact that mobile phones differ from desktop computers since mobile phones are not capable of performing the demanding processing required as in emotion recognition. To surpass this fact, in our approach, mobile phones are required to transmit all data collected to a server which is responsible for performing, among other, emotion recognition. The object oriented architecture that we have created, combines evidence from multiple modalities of interaction, namely the mobile device's keyboard and the mobile device's microphone, as well as data from user stereotypes. All collected information is classified into well-structured objects which have their own properties and methods. The resulting emotion detection platform is capable of processing and re-transmitting information from different mobile sources of multimodal data during human–computer interaction. The interface that has been used as a test bed for the affective mobile interaction is that of an educational m-learning application.

8.1 Overview of the Emotion Recognition System's Architecture

The authors of Neerincx and Streefkerk (2003) describe a study where emotion, trust and task performance are investigated as important elements of user interaction with mobile services. The participants of this study performed interaction tasks with mobile services, using small handheld devices and laptops. This study concludes with the presentation of the relations between trust, performance, devices and emotions of the users. In Gee et al. (2005) mobile telephones were used to collect data in order to find the relationship between gambling and mood state from gamblers. The results of this study revealed that subjective anxiety/arousal levels

E. Alepis and M. Virvou, *Object-Oriented User Interfaces for Personalized Mobile Learning*, Intelligent Systems Reference Library 64,
DOI: 10.1007/978-3-642-53851-3_8, © Springer-Verlag Berlin Heidelberg 2014

were significantly higher during and after gambling than during the urge to gamble. The authors of this study also state that collecting data through the use of mobile telephones appeared to be a valuable development in their research. Isomursu et al. (2007) collect affective interaction data that emerge from mobile applications using several emotion collection methods. As a next step, these methods are evaluated and the authors discuss about the experiences they have gained using these methods, and also provide a comparison framework to summarize their results.

These studies provided strong evidence that emotion recognition in mobile interaction is very important, though it is difficult to be achieved. In this section, we describe the general architecture of the out emotion recognition system. Data related to the general emotion recognition process are divided into two individual categories. The first category comprises of the multimodal data that may be collected by each individual modality of interaction. The emotion detection server supports one or more modalities and each of them are factually the emotion detection server's components. Correspondingly, emotional states are object attributes and methods of each modality.

In the implementation of the system, there is a set of the available modalities. The authors have presented in previous work (Tsihrintzis et al. 2008) emotion recognition systems that consist of multiple modalities. As it is illustrated in Fig. 8.1, the mobile server may manipulate information by its available mobile clients. Each client has a number of services available, after its connection with the main server. Correspondingly, each modality may provide the system with information about the recognition of one or more emotional states with specific degrees of certainty, along with information concerning the user action attributes and the triggering conditions. Additionally, each modality may also provide supplementary information concerning the emotional states of users, which are associated with specific user actions during the human–computer interaction. Such actions include correct or wrong browsing, answers in tests in educational software environments, etc.

For the application of OO model into affective e-learning interfaces, stereotypic information has also been used. A complete study considering the incorporation of stereotypes into emotion recognition systems is beyond the scopes of this chapter and is shown in previous work of the authors (Alepis and Virvou 2006). However, in this chapter we suggest an object oriented structure of all the available stereotypic information. As it is illustrated in Fig. 8.2, the main user model stereotype class is associated with four subclasses. The first class stores and administers data that are associated with users' characteristics. These characteristics derive from each user's personality, such as the user's educational level, the user's sex and age, gender, etc. and help the system improve its emotion recognition capabilities. The second subclass models stereotypic information concerning user actions during their interaction with personal computers, while the third subclass represents prestored information about each modality's ability of recognizing each one of the available emotional states.

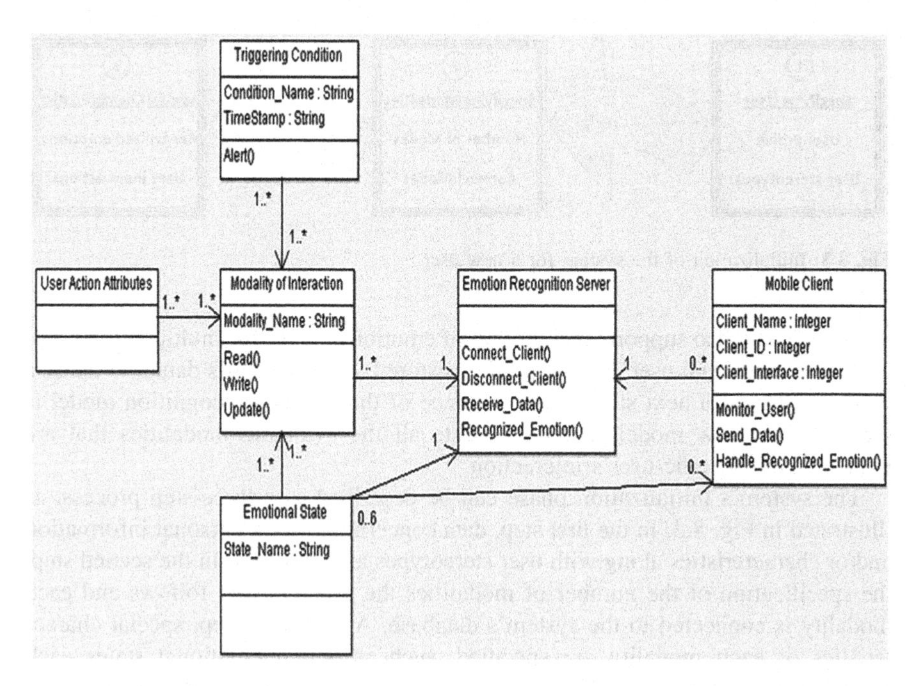

Fig. 8.1 Class model for the emotion recognition server

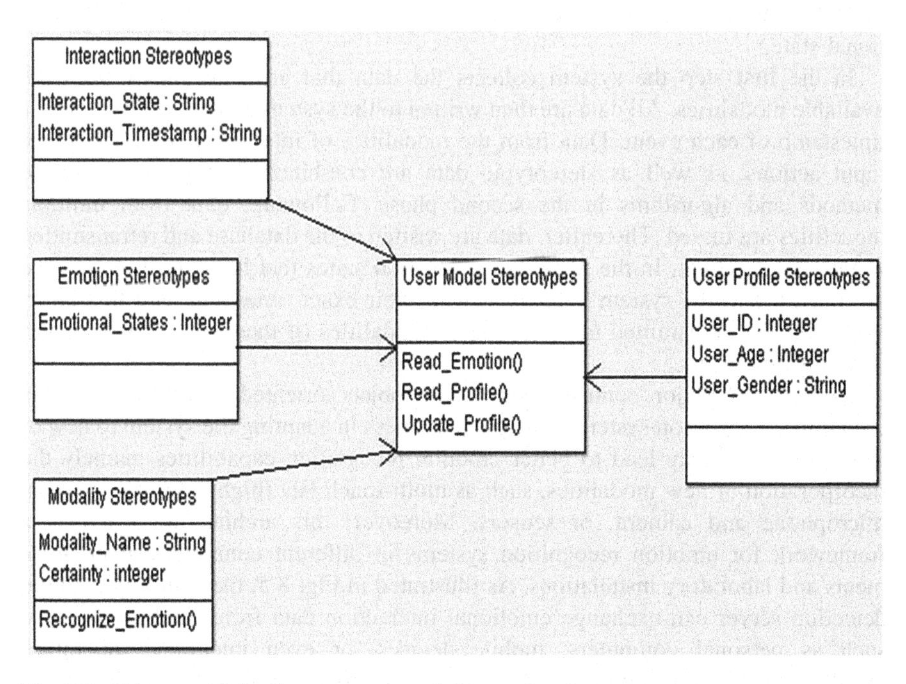

Fig. 8.2 Object model for the construction of the emotional stereotypes

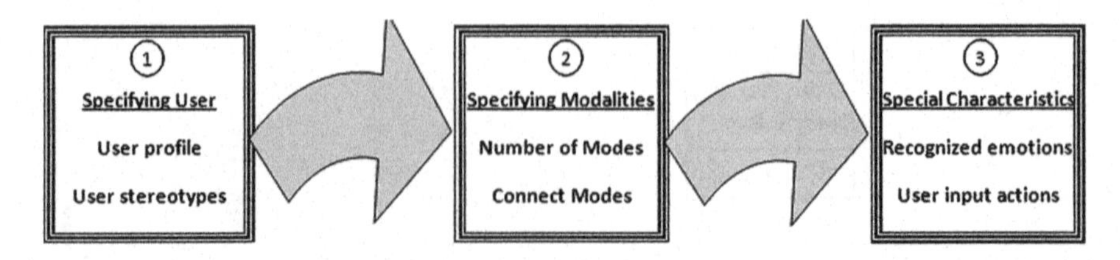

Fig. 8.3 Initialization of the system for a new user

The system also supports recognition of emotional states for multiple users. For each new individual user, his/her profile is stored to the system's database under a new user_id. As a next step, a new instance of the emotion recognition model is created. This new model will incorporate all the available modalities that are related to the specific user's interaction.

The system's initialization phase can be described as a three-step process, as illustrated in Fig. 8.3. In the first step, data concerning user's personal information and/or characteristics along with user stereotypes are collected. In the second step, the specification of the number of modalities the user is using follows and each modality is connected to the system's database. At the final step, special characteristics of each modality are specified, such as which emotional states each modality supports and which user input actions may be added additionally.

Correspondingly, Fig. 8.4 illustrates the system's five-step process of the available multimodal emotional information in order to recognize a user's emotional state.

In the first step the system collects the data that are transmitted from the available modalities. All data are then written to the system's main database with a timestamp of each event. Data from the modalities of interaction, data from user input actions, as well as stereotypic data are combined through sophisticated methods and algorithms in the second phase. Following, data from multiple modalities are fussed. Thereafter, data are written to the database and retransmitted to the mobile client. In the final step, emotional states that have been recognized are recorded to the system's database with their exact timestamp and this information is also transmitted to the available modalities (if these modalities support this transmission).

One of the major contributions of the object oriented architecture of the emotional recognition system is the great easiness in adapting the system to new or more roles that may lead to better emotion recognition capabilities, namely the incorporation of new modalities, such as multi-touch HD (high definition) screen, microphone and camera, or sensors. Moreover, this architecture provides a framework for emotion recognition systems in different computerized environments and laboratory installations. As illustrated in Fig. 8.5, the resulting emotion detection server can exchange emotional interaction data from multiple sources, such as personal computers, mobile devices, or even integrated laboratory installations with multiple devices as input modalities. It also gathers information

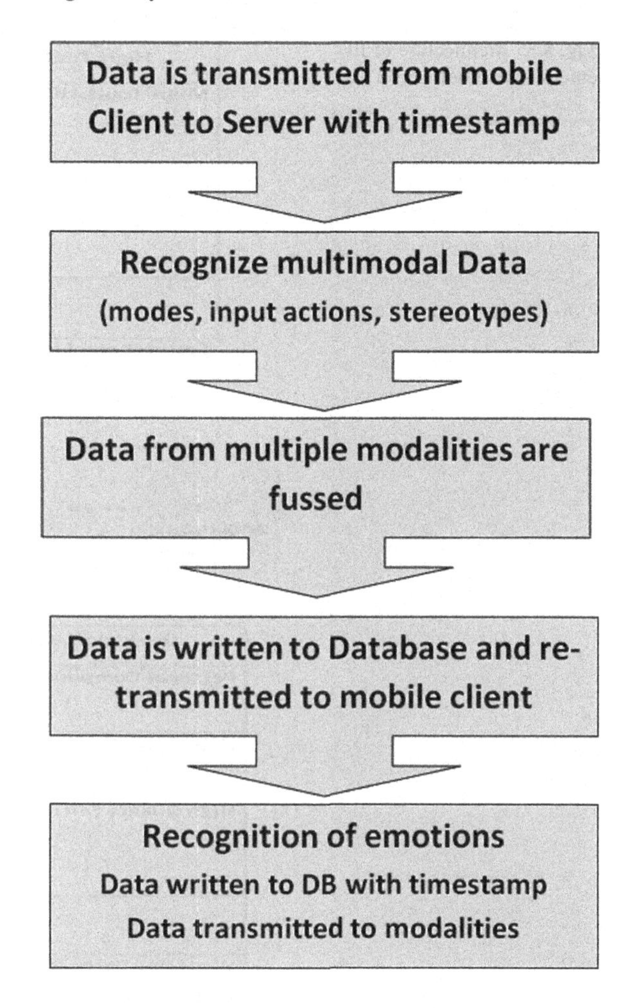

Fig. 8.4 Processing multimodal data for emotion recognition

from the user monitoring database and the application database. However, in this chapter we focus on the integration of mobile devices, as well as on the usefulness of such integrations according to the OO structured model.

8.2 Emotion Recognition Data into Objects

In Sect. 8.3, we have shown how the object oriented method can be used in order to provide a reliable model that stores and handles the information used in emotion recognition systems. Our expectation is that this approach will be adopted in future emotion recognition systems with more modalities and improved emotion detection mechanisms. In this section we present actual emotion recognition data

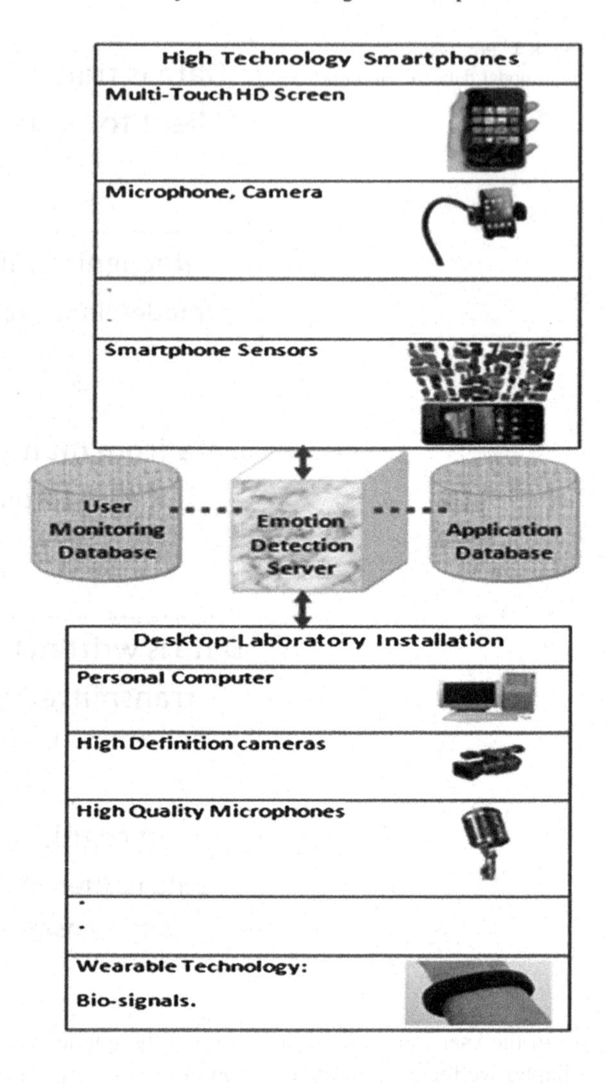

Fig. 8.5 Architecture of the emotion detection server

resulting from a bi-modal mobile educational application that are classified according to the aforementioned object oriented architecture. The available information (users' actions, users' stereotypic characteristics and data from the available modalities) are pre-processed and structured using the OO model and then transmitted to the main server. The aim of this section is to illustrate the variety of different data in order to indicate the necessity of a well structured approach which can classify and manipulate them. In this study we have developed a prototype emotion detection server with respect to affect recognition concerning six basic emotion states. These emotion states are happiness, sadness, surprise, anger and disgust as well as the emotionless state which we refer to as neutral.

In order to classify all the aforementioned emotion recognition data we have used two basic theories, as mechanisms for the emotion recognition process, namely the Artificial Neural Networks (ANNs) theory and the Multi Attribute Decision Making (MADM) theory. In these approaches, which are thoroughly described in Virvou et al. (2012), we produce a six-dimensional output vector which can be regarded as the degree of membership of the bi-modal information in each of the "neutral", "happiness", "surprise", "anger", "disgust" and "sadness" classes. The overall functionality of these approaches which lead to multi-modal recognition of emotions through multimodal data, described in Virvou et al. (2012), requires more comprehensive writing and is beyond the scope of the present chapter. In this chapter, we focus specifically on providing an efficient and flexible structure for storing and handling the available bi-modal information as well as information about user characteristics concerning emotion recognition in order to improve our system's emotion recognition capabilities. Furthermore, the proposed structure can further improve the algorithms' accuracy, since the available information considering the users' affective interaction is collected from multiple sources (multiple mobile devices operated by different users) and is finally stored in the central emotion detection server. This derives from the fact that in both the algorithms that we have used there is a considerable improvement in their emotion detection accuracy when the amount of the available data they are able to process increases.

8.3 Overview of the Mobile System

The basic architecture of the mobile bi-modal emotion recognition subsystem is illustrated in Fig. 8.6. Participants were asked to use their mobile device and interact with a pre-installed educational application. Their interaction could be accomplished either orally (through the mobile device's microphone) or by using the mobile devices keyboard and of course by combining these two modes of interaction. All data were captured during the interaction of the users with the mobile device through the two modes of interaction and then transmitted wirelessly to the main server. More specifically, a user monitoring component has been used to capture all user input actions and then pass the information to the educational application. Correspondingly, all input data from the captured actions and data from the educational application (such as correct or wrong answers to test, navigation errors and spoken words unrelated to the educational application) are transmitted wirelessly to the Emotion Detection Server through the use of mobile cloud services, where emotion detection processes take place and finally the data are stored in the system's database. All the input actions were used as trigger conditions for emotion recognition by the emotion detection server. Finally, all input actions as well as the possible recognized emotional states were modeled according to the OO approach and were also stored in the system's database. The discrimination between the participants is done by the application that uses the

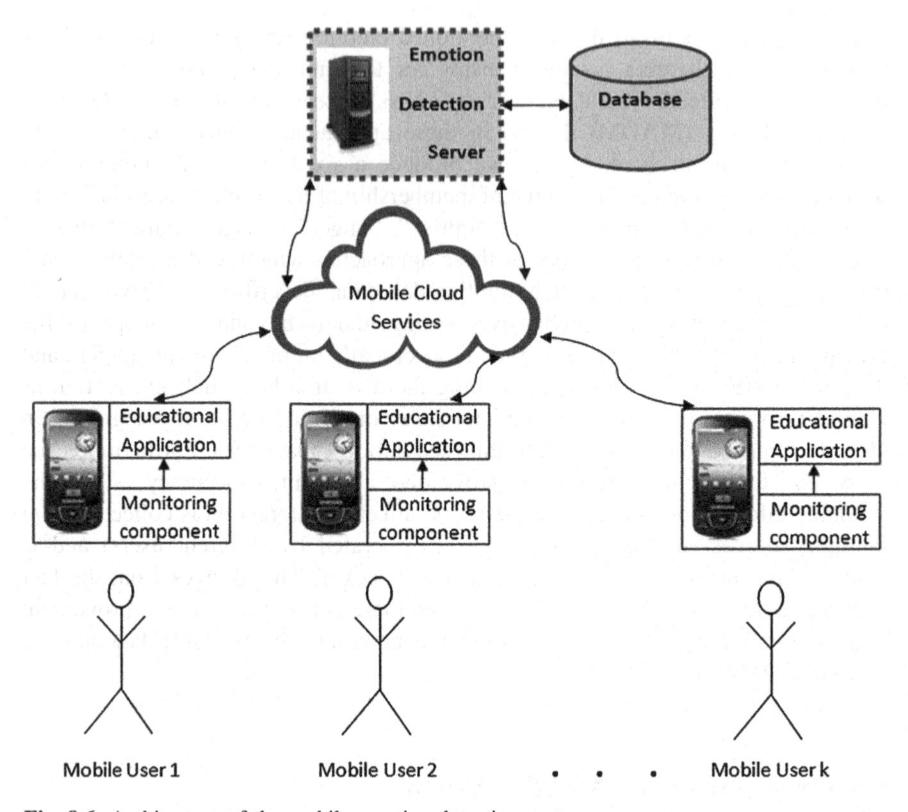

Fig. 8.6 Architecture of the mobile emotion detection system

main server's database and for each different user a personal profile is created and stored in the database. In order to accomplish that, username and password are always required to gain access to the mobile educational application.

A snapshot of a mobile emulator, operated by a user who participated in the evaluation study is illustrated in Fig. 8.7. Participants were prompted to answer questions and also to take tests by using a mobile device. They were able to write their answers through the mobile device's keyboard, or alternatively give their answers orally, using their mobile device's microphone. When participants answered questions, the system also had to perform error diagnosis in cases where the participants' answers have been incorrect.

8.4 Data Associated with User Characteristics

As shown in Virvou et al. (2007), different computer modalities have distinguishable capabilities in recognizing humans' emotional states. For example, some emotions may give significant visual evidence of their existence to any observer,

Fig. 8.7 A user is answering a question orally

while in other situations audio-lingual data may be preferable. People who use mobile devices have also a tendency to express themselves emotionally while interacting with such devices. This fact is further emphasized in cases where mobile devices are used as an instrument of communication with other people. Correspondingly, in human-mobile interaction, specific categories of users may use different modalities for their interaction. In an empirical study that we conducted in previous research work, we have noticed that a very high percentage (85 %) of young people (below 30 years old), who are also inexperienced with personal computers, reported to have preferred expressing themselves through the oral mode rather than the keyboard during their interaction with a computer. On the contrary, participants who were computer experts did not give us considerable data for the affect perception during the oral communication with their computer. In our approach, the categorization of users has been done based on their age, their gender, their computer knowledge level, as well as their educational knowledge level. Categorizing our users has helped us built the corresponding user models as well as stereotypes, that are described in Sect. 4.4. This categorization emerges from past empirical studies that the authors have conducted for emotion recognition purposes (Virvou et al. 2007).

An important issue concerning the combination of data from multiple modalities are also the fact that different modalities may give evidence for different emotions or emotional states. A modality may be able to provide a system with information about a number of discrete emotions, while another modality may only decide whether positive or negative feelings are detected. Correspondingly, specific characteristics of a user may lead the system to focus on a specific modality that this user may prefer to use for his/her interaction.

8.5 Data from User Input Actions

The emotion recognition system incorporated a user monitoring component that captured all users' actions concerning two modalities of interaction, namely the mobile device's keyboard and the mobile device's microphone. After processing the collected data we also came up with statistical results that associated user input actions through the keyboard and microphone with emotional states. More specifically, considering the keyboard, we have the following categories of user actions: (k1) user types normally (k2) user types quickly (speed higher than the usual speed of the particular user) (k3) user types slowly (speed lower than the usual speed of the particular user) (k4) user uses the "delete" key of his/her personal mobile device often (k5) user presses unrelated keys on the keyboard (k6) user does not use the keyboard. These actions were also considered as criteria for the evaluation of emotion recognition with respect to the user's actions through the keyboard.

Considering the users' basic input actions through the mobile device's microphone we come up with seven cases: (m1) user speaks using strong language (m2) users uses exclamations (m3) user speaks with a high voice volume (higher than the average recorded level) (m4) user speaks with a low voice volume (low than the average recorded level) (m5) user speaks in a normal voice volume (m6) user speaks words from a specific list of words showing an emotion (m7) user does not say anything. These seven actions were also considered as criteria for the evaluation of emotion recognition with respect to what users say (paralinguistic information) and how they say it (linguistic information).

All user input actions are handled as objects where an action may occur with a specific degree of certainty and a timestamp that indicates when this action occurred. The corresponding subclass for user input actions is illustrated in Fig. 8.1.

8.6 Stereotypic Data Analysis

Considering the critical problem in affective computing as to which mode gives better results or to what extent should the evidence from each mode be taken into account, the authors have proposed in the past a novel approach for calculating weights of significance for each modality based on stereotypes and a multi-criteria theory (Alepis and Virvou 2006; Virvou et al. 2007). The incorporation of stereotypes into emotion recognition systems led the systems to improve their emotion recognition accuracy. Thus, this approach is also incorporated into the OO model. The object model for emotion stereotypes is illustrated in Fig. 8.2. Stereotype-based reasoning takes an initial impression of the user and uses this to build a user model based on default assumptions (Kay 2000). Stereotypes constitute a powerful mechanism for building user models (Kay 2000). This is due to the fact that stereotypes represent information that enables the system to make a large number of plausible inferences on the basis of a substantially smaller number of observations (Rich 1983). The stereotype inferences are used in combination with a decision theory, namely Simple Additive Weighting (SAW) (Fishburn 1967; Hwang and Yoon 1981) for estimating weight of significance of each mode in the affective reasoning of the system for a particular user.

In previous research work of the authors (Alepis and Virvou 2006), we have classified our users into stereotypes concerning their age, their educational level, their computer knowledge level and their gender. A four-dimensional stereotypic vector of the form:

(User_Name, Stereotypic Characteristic1, Stereotypic Characteristic2, Stereotypic Characteristic3, Stereotypic Characteristic4), is used to assign each user model to a stereotypic category that can provide additional emotion recognition information.

Stereotypic Characteristic1 refers to the user's age. *Stereotypic Characteristic 2* refers to the user's computer knowledge level. Similarly *stereotypic characteristics 3* and *4* refer to the user's educational level and to the user's gender respectively. The inferences of this stereotype vector provide information about the weights of importance of each mode for the users belonging to that stereotype. For example, stereotypes give us the important information that younger mobile device users have a statistical tendency to express their feelings through the oral mode of interaction, in comparison with older mobile device users who do not tend to express their emotions orally.

Emotional stereotypes can provide inferences concerning hypothesis about users' feelings and which modality should be more important for providing evidence about users' feelings. More specifically, in many cases, data from either the vocal mobile interaction or the interaction through the mobile keyboard give evidence of different emotions with quite similar degrees of certainty. For example, the system may have evidence that a user is either angry, while saying or typing something, or stressed, or even confused. The incorporation of stereotypes in the system provides inferences concerning people belonging to the same category with the user that may help in recognizing an emotion that is more common for the users of this category among others. Evidence for the character or the personality of a particular user may raise the degree of certainty for a particular emotion recognized.

8.7 Conclusions

In this chapter, we described a multimodal emotion recognition system that is structured according to the object oriented method. The system uses the OO approach that combines evidence from multiple modalities of interaction and data from emotion stereotypes and classifies them into well structured objects with their own properties and methods. Advantages of the proposed approach include the well-known conveniences and capabilities of object oriented structures, such as easiness in the system's maintenance, great extensibility, better communication through different modalities, good cooperation with different object oriented programming languages, easiness in code debugging, as well as code reusability.

The present system's architecture can be adopted in future emotion recognition systems with multiple modalities and improved emotion detection algorithms. Furthermore, the resulting emotion detection server is capable of using and handling transmitted information from different sources of human–computer interaction. Independent user interfaces may send wirelessly or wired information about users' interaction to the emotion detection server and the server can respond with information about possibly recognized emotional states of the users.

The results of the system's evaluation study are presented in the next chapter and have been very positive and promising. As for future work, we plan to extend our evaluation studies with more experiments of students as users of the system and also exploit more sources of data on modalities of affective interaction.

References

Alepis E, Virvou M (2006) Emotional intelligence: constructing user stereotypes for affective bi-modal interaction. In: Knowledge-based intelligent information and engineering systems 2006. Lecture notes in computer science LNAI-I, vol 4251. Springer, Heidelberg, pp 435–442

Fishburn PC (1967) Additive utilities with incomplete product set: applications to priorities and assignments. Oper Res 15(3):537

Gee P, Coventry KR, Birkenhead D (2005) Mood state and gambling: using mobile telephones to track emotions. Br J Psychol 96(1):53–66

Hwang CL, Yoon K (1981) Multiple attribute decision making: methods and applications. Lecture notes in economics and mathematical systems, vol 186. Springer, Heidelberg

Isomursu M, Tähti M, Väinämö S, Kuutti K (2007) Experimental evaluation of five methods for collecting emotions in field settings with mobile applications. Int J Hum Comput Stud 65(4):404–418

Kay J (2000) Stereotypes, student models and scrutability. In: Gauthier G, Frasson C, VanLehn K (eds) Proceedings of the 5th international conference on intelligent tutoring systems. Lecture notes in computer science, vol 1839. Springer, Heidelberg, pp 19–30

Neerincx M, Streefkerk JW (2003) Interacting in desktop and mobile context: emotion, trust, and task performance. Lecture notes in computer science (Lecture notes in artificial intelligence and Lecture notes in bioinformatics), vol 2875. pp 119–132

Rich E (1983) Users are individuals: individualizing user models. Int J Man Mach Stud 18:199–214

Tsihrintzis G, Virvou M, Stathopoulou IO, Alepis E (2008) On improving visual-facial emotion recognition with audio-lingual and keyboard stroke pattern information, Web intelligence and intelligent agent technology, WI-IAT'08, vol 1. pp 810–816

Virvou M, Tsihrintzis G, Alepis E, Stathopoulou IO, Kabassi K (2007) Combining empirical studies of audio-lingual and visual-facial modalities for emotion recognition. In: Knowledge-based intelligent information and engineering systems—KES 2007. Lecture notes in computer science (Lecture notes in artificial intelligence), vol 4693/2007. Springer, Berlin, pp 1130–1137

Virvou M, Tsihrintzis GA, Alepis E, Stathopoulou I-O, Kabassi K (2012) Emotion recognition: empirical studies towards the combination of audio-lingual and visual-facial modalities through multi-attribute decision making. Int J Artif Intell Tools 21(2) (Art. no 1240001)

Chapter 9
Evaluation of the Multimodal Object Oriented Architecture

Abstract This chapter describes an evaluation study for an application of the Object Oriented architecture of a multimodal mobile system. A system relying on this structure is described in the previous chapter. In this chapter the authors evaluate the "quality" of their approach by attempting to provide solutions to the problems of successfully handling multimodal data in the much demanding area of mobile affective interaction. The results in this chapter's findings indicate the success of their project and also strengthens their belief that the OO paradigm can successfully handle mobile multimodal data.

9.1 Evaluation Study

The resulting system that was presented in the previous chapter, may be considered successful if it is approved by scientists that use such systems and is also beneficial to software users. For this reason an evaluation study was conducted among 15 computer scientists of an informatics department and 15 software programmers. Both scientists and programmers had previous experience in the field of affective computing. All participants attended a short presentation of the capabilities of the emotion recognition server prototype, during which the servers' architecture and design was analyzed. Subsequent to the presentation, we demonstrated the whole process while having interconnections between mobile devices and the emotion recognition server. Participants were also asked to use the educational programming platform that was pre-installed in a smartphone that was given to them and the whole interaction was videotaped. Considering the software that ran on the emotion recognition server we have used Java as an object oriented language, while for the implementation of the mobile devices' applications we have used the Java in conjunction with the Android Software Development Toolkit (SDK). The mobile devices also incorporated pre-installed software that recorded all user actions and transmitted the data to the servers' database for emotion recognition analysis. This software was designed to run "silently" in the background of the mobile operating

E. Alepis and M. Virvou, *Object-Oriented User Interfaces for Personalized Mobile Learning*, Intelligent Systems Reference Library 64,
DOI: 10.1007/978-3-642-53851-3_9, © Springer-Verlag Berlin Heidelberg 2014

system and was meant to capture all kinds of user input actions that could be used for emotion recognition. These actions include typing on a mobile device's keyboard, audio data from microphones and optical data from mobile front cameras. At the end of the study we also presented to our audience how other types of data, such as bio-signals from wearable mobile devices, could also be incorporated and modeled into objects that could be easily stored in the emotion recognition server.

An important issue considering our evaluation study was the fact that we had to ensure that we could elicit emotions of our participants so that these could be recorded in the system's database for further analysis and use. The elicitation of users' emotions constitutes an important part of the settings of studies concerning affective computing. For example, Nasoz and Lisetti (2006) conducted an experiment where they elicited six emotions (sadness, anger, surprise, fear, frustration and amusement) and measured physiological signals of subjects by showing them movie clips from different movies (The Champ for sadness, Schindler's List for anger, The Shining for fear, Capricorn One for surprise, and Drop Dead Fred for Amusement) and by asking them hard mathematical questions (for frustration). They conducted a panel study to choose an appropriate movie clip to elicit each emotion. In contrast, our study aimed at recording emotions for different modalities; we did not aim at recording physiological signals of users but rather aimed at recording and analyzing keyboard-stroke and audio-lingual expressions of emotions. For the purposes of our studies, we used an educational application to elicit the emotions of the participants.

Indeed, for the elicitation of the five basic emotions of our study (happiness, anger, sadness, surprise and neutral emotional state) the educational application provided the following situations:

1. For happiness: an animated tutoring agent would say jokes and would make funny faces or the user would receive excellent grades in his/her performance on tests.
2. For anger: the animated tutoring agent would be rude and unfair to users for their performance at tests.
3. For sadness: the animated agent would look sad in cases where the user's performance was below the expected one.
4. For surprise: the animated tutoring agent would pop up into the screen completely unexpectedly.
5. For neutral: the user would use the educational application under normal circumstances.

Mobile devices were given to all 30 participants of the evaluation study. The entire interaction for each one of the participants was also video recorded by an external video camera. After completing the interaction with their mobile device, participants were asked to watch the video clips concerning exclusively their personal interaction and to determine in which situations they experienced changes in their emotional state. Correspondingly, these changes were associated with one of the five basic emotional states that the system tries to recognize. To this end, all

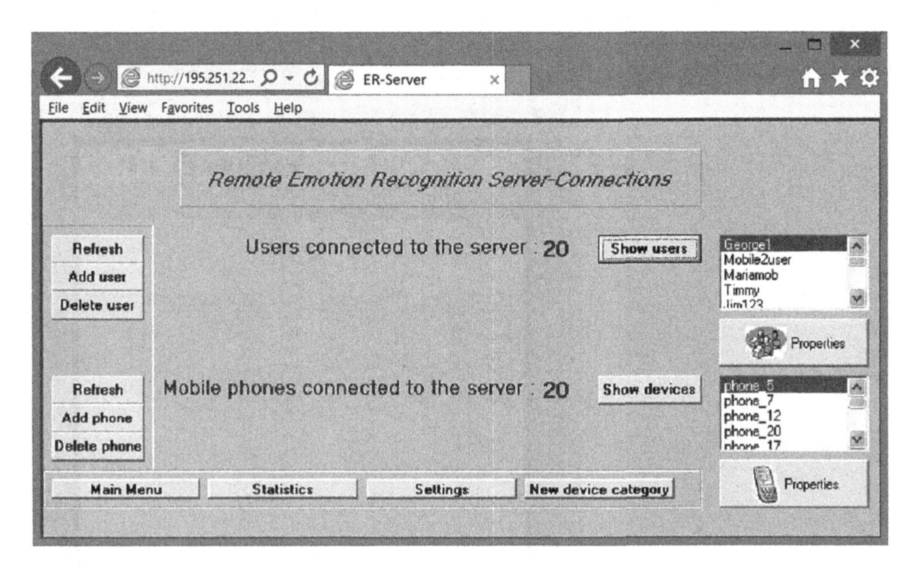

Fig. 9.1 Users and mobile devices connected to the server

the data was also recorded and time stamped. In this way, we managed to compare the system's assumptions about the users' recognized emotions and the actual emotions of the users.

Figure 9.1 illustrates a snapshot of the users and the mobile devices that were connected to the web interface of the remote emotion recognition server. Figure 9.2 illustrates emotions that had been recognized by the application after the processing of the available multimodal interaction data. When a trigger condition occurs, the systems makes a hypothesis about the recognition of an actual emotional state by choosing the most evident of the six available emotional states. The prevalent emotional state is illustrated in Fig. 9.2 as the emotional state with the highest value in each record.

At the end of the evaluation study, a part of the emotion detection servers' processed data, which was recorded during the participant's interaction with the mobile devices, was presented for further discussion. Each participant could see what was recorded during his/her interaction with the mobile device, as well as the system's assumptions about his/her emotional state at each moment. Figures 9.3, 9.4 and 9.5 illustrate representative results of the evaluation study considering the participants' opinions. More specifically, Fig. 9.3 presents the participants' opinions on the usefulness level of the emotion detection servers' architecture. Figure 9.4 presents the participants' opinions for the system's capabilities in using and handling multimodal data from multiple sources. In Fig. 9.5 the participants rated the systems capabilities that derive from its object oriented architecture. As can be noted in Fig. 9.5, not including platform independence all remaining participants' ratings stand above 70 %. Providing full platform independence for mobile devices is something quite difficult in the current era of mobile technology,

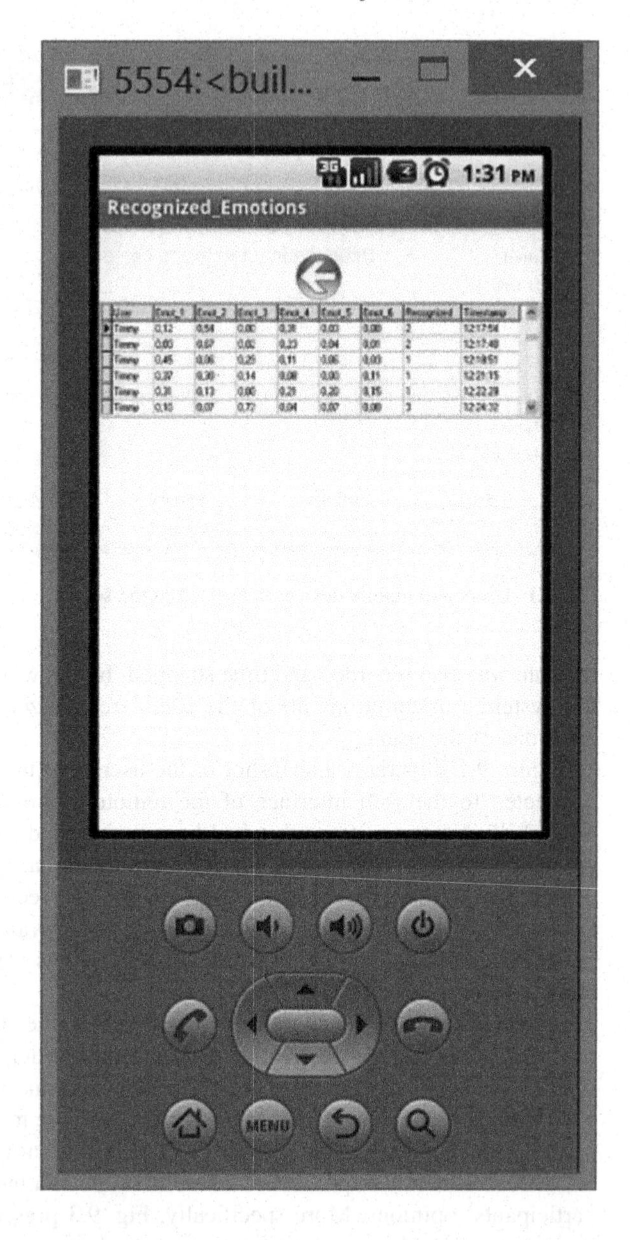

Fig. 9.2 Emotion detection data log for a particular user

where technological changes and updates are a daily phenomenon, since mobile programming needs a number of steps to reach maturity and compare with classical desktop application programming.

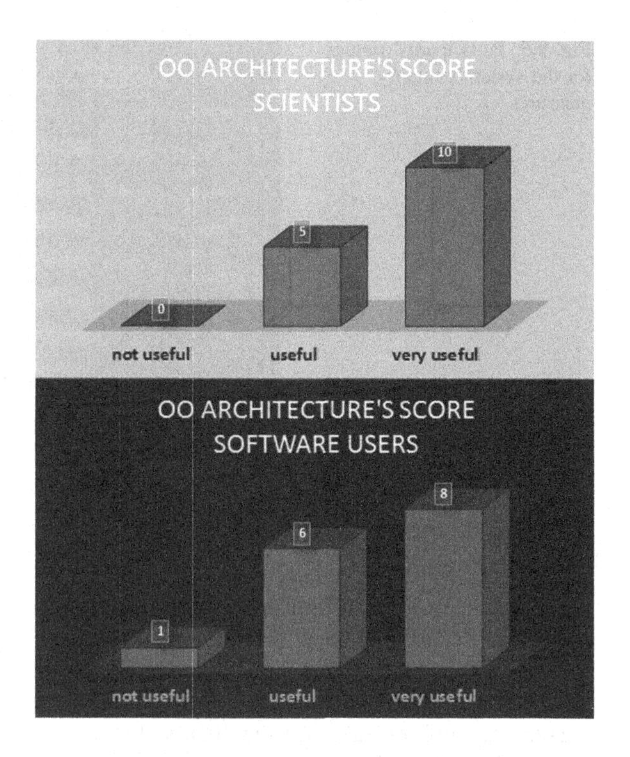

Fig. 9.3 Usefulness of the system's architecture

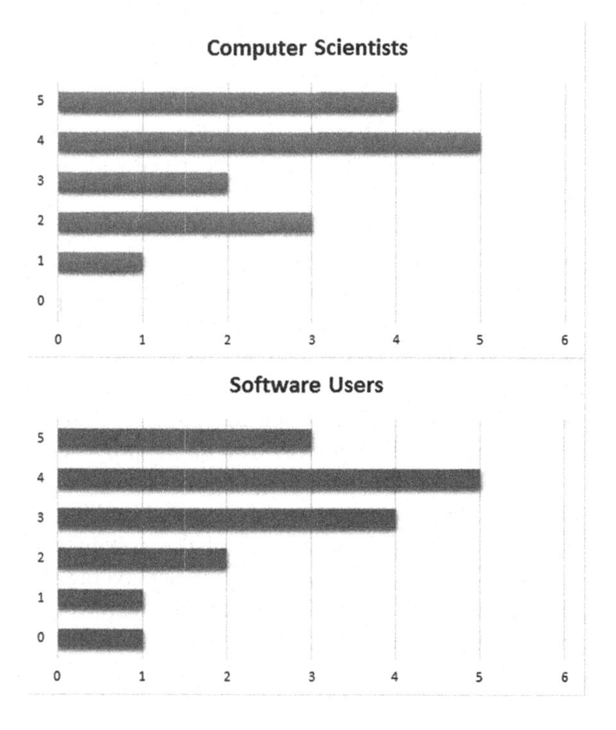

Fig. 9.4 Ranking the system's capabilities in successfully handling multimodal information

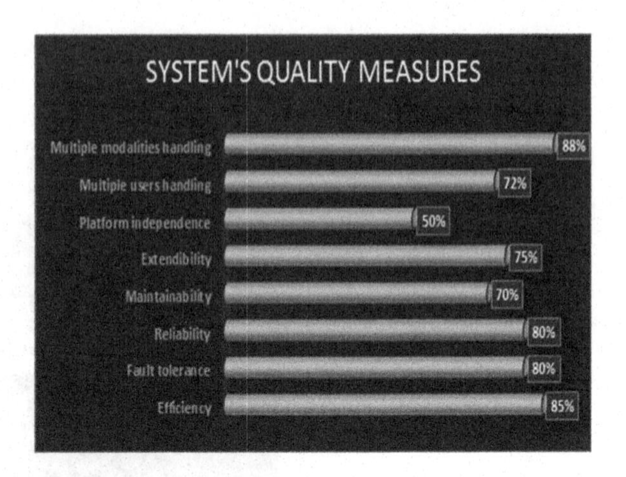

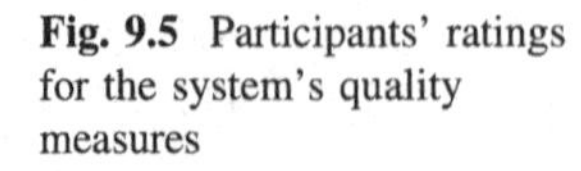
Fig. 9.5 Participants' ratings for the system's quality measures

9.2 Discussion

As stated in (Pantic and Rothkrantz 2003) and in (Picard 2003), there is no readily accessible database of tests on multimodal material that could be used as a basis for benchmarks for efforts in the research area of automated human affective feedback analysis. As a result, we cannot conclude that a system achieving a high average recognition rate performs better than a system attaining a lower average recognition rate, unless we can assure the same evaluation settings and the same emotion recognition databases. For this purpose, we compared our system's results with previous work by the authors concerning mobile affective interaction and also human computer affective interaction. We should also emphasize that the effort to recognize emotional states in mobile user interfaces, as described in previous chapters, is very recent and has not derived with any sufficient scientific results and conclusions.

Table 9.1 illustrates successful emotion recognition by comparing two affective mobile approaches. The first is the standard emotion recognition approach that has been evaluated with the use of mobile devices in previous work by the authors (Alepis et al. 2008). The second is this chapter's approach that combines multimodal evidence from the available modalities and also data from user stereotypes and user input actions. The OO architecture presents considerable improvements in the resulting system's emotion recognition accuracy in comparison with emotion recognition in mobile devices without the proposed architecture. From Table 9.1 it is also apparent that the average correct recognition rate (for all the six emotional states) of the system with our proposed architecture stands at 68 %, while the average correct recognition rate of the system without the use of the proposed architecture is 59 %. There is a significant average 9 % improvement in the mobile system's emotion recognition accuracy which results mainly from the fact that the OO architecture allows for a better usage of all the available information that is related to emotion recognition during human-mobile device interaction. We should

Table 9.1 Emotion recognition rates in mobile applications with and without the OO architecture

Emotions	Recognition in a standard mobile application	Recognition in mobile application with OO architecture
Neutral	0.46	0.60
Happiness	0.64	0.80
Sadness	0.70	0.75
Surprize	0.45	0.61
Anger	0.70	0.73

Table 9.2 Emotion recognition rates in desktop (bi-modal and tri-modal) and mobile user interfaces

Emotions	Desktop tri-modal recognition	Desktop bi-modal recognition	Mobile bi-modal recognition
Neutral	0.95	0.73	0.60
Happiness	0.92	0.74	0.80
Sadness	0.85	0.80	0.75
Surprize	0.90	0.67	0.61
Anger	0.82	0.80	0.73

emphasize that in both mobile interactions the algorithmic approaches that were used for the recognition of emotions were the same. Furthermore, Table 9.1 gives us the only factual and reliable comparison between the two existing affective mobile approaches, since, as it is already mentioned, there are no existing results in the scientific literature regarding emotion recognition in mobile devices to date.

The authors also compared their results of emotion recognition in mobile devices with results regarding emotion recognition in desktop computers (Tsihrintzis et al. 2008), (Stathopoulou et al. 2010). For the emotion recognition in desktop computers we analyzed the results from a system with two modalities of interaction (microphone and keyboard) and also a system with three modalities of interaction (microphone, keyboard and camera). The results of these comparisons are illustrated in Table 9.2.

Although emotion recognition is still essentially better in the HCI rather than in the interaction between users and their mobile phones, we should state that if we compare only the bi-modal systems (desktop and mobile) their average accuracy levels differ only by small rates (the desktop interaction outperforms the mobile interaction). It is the authors' opinion that this is not because of the difference of the devices processing power, since the emotion recognition process using mobile phones in our study takes place in a remote server. We believe that this minor limitation in the mobile system's emotion recognition accuracy, in comparison with an equivalent desktop system, derives from the limited capabilities that the mobile devices have when considered as input and output devices. As we have stated in Sect. 9.1, desktop computers incorporate better data input devices than mobile phones (keyboard, microphone) and also have higher resolution monitors (as data output devices) and better quality speakers that also help in the fore mentioned elicitation of human emotions during human–computer interaction.

The incorporation of a third modality (camera) provides significant improvements in the system's emotion recognition capabilities and this is the main reason why the tri-modal desktop emotion recognition system of Table 9.2 outperforms the other approaches. On the whole, the visual modality has been found to play a very important role in the successful recognition of human emotions through computers (Stathopoulou et al. 2010).

9.3 Conclusions

Computer scientists and software users appreciated both the system's proposed architecture and its capabilities in using and handling transmitted information from different sources of data during human–computer interaction. The emotion detection server has been considered as very user-friendly by all scientists and programmers with previous comprehensive experience in the field of affective computing. Furthermore, the resulting affective mobile interaction system with the OO architecture outperforms the affective mobile interaction system without the OO architecture in terms of emotion recognition capabilities. However, the participants suggested that there should be a further and more extended evaluation study for long term users working with the educational software, such as students, so that the existing long term and short term user models could be further improved. They also stated that in the future, new devices and modalities would offer more sophisticated and comprehensive emotion recognition abilities.

References

Alepis E, Virvou M, Kabassi K (2008) Knowledge engineering for affective bi-modal interaction in mobile devices. In: Knowledge-based software engineering. Frontiers in artificial intelligence and applications, vol 180. IOS Press, Amsterdam, pp 305–314, ISBN 978-1-58603-900-4

Nasoz F, Lisetti CL (2006) MAUI avatars: mirroring the user's sensed emotions via ex-pressive multi-ethnic facial avatars. J Vis Lang Comput 17:430–444

Pantic M, Rothkrantz LJM (2003) Toward an affect-sensitive multimodal human-computer interaction. Proc IEEE 91(9):1370–1390

Picard RW (2003) Affective computing: challenges. Int J Hum Comput Stud 59(1–2):55–64

Stathopoulou I-O, Alepis E, Tsihrintzis GA, Virvou M (2010) On assisting a visual-facial affect recognition system with keyboard-stroke pattern information. Knowl-Based Syst 23(4):350–356

Tsihrintzis G, Virvou M, Stathopoulou IO, Alepis E (2008) On improving visual-facial emotion recognition with audio-lingual and keyboard stroke pattern information, web intelligence and intelligent agent technology, WI-IAT'08, vol 1, pp 810–816

Chapter 10
Mobile Affective Education

Abstract This chapter introduces a new programming language for children named m-AFOL. This programming language extends a "desktop" version of a programming language named AFOL (Alepis 2011), already developed by one of the authors. As with AFOL, the m-AFOL programming language has been based on the idea of the well-known Logo programming language. However, m-AFOL extends Logo's basic programming concepts such as sequential and functional programming by introducing the more modern concepts of Object Oriented programming. Furthermore, m-AFOL incorporates highly sophisticated user interaction mechanisms, namely affective interaction through emotion recognition and through the use of animated tutoring agents. Perhaps the most important addition to this programming language is the mobile interface with the platform. Through a mobile application, pre-installed on a modern smartphone with a wireless internet connection, a highly sophisticated graphical user interface with affective interaction capabilities is transformed into a modern programming language learning tool for children.

10.1 Background

The well-known "Logo" programming language was introduced in 1967 (Frazier 1967). The Logo developers' main objective was to take the best practices and ideas from computer science and computer programming and produce an interface that was good and suitable for the education of young children. Hence, the authors of Logo aimed to create a friendly programming language for the education of children where they could learn programming while having fun by playing with words and sentences. The first implementation of Logo was written in LISP programming language for the purposes of creating a programming language as a math land where kids could play by giving commands that produced nice and colorful drawings. Logo programming language may be seen as a compromise between a sequential programming language with block structures, and a functional programming

E. Alepis and M. Virvou, *Object-Oriented User Interfaces for Personalized Mobile Learning*, Intelligent Systems Reference Library 64,
DOI: 10.1007/978-3-642-53851-3_10, © Springer-Verlag Berlin Heidelberg 2014

language. Logo has been used mainly in the past as a teaching language for children but its list handling facilities made it remarkably useful for producing useful scripts. A detailed study on the "Logo" programming language from its early stages and also recent work on Logo-derived languages and learning applications can be found in Feurzeig (2010).

Modern programming languages try to provide as much user-friendliness as possible while retaining their full programming functionality. Hudlicka (2003) points out that an unprecedented growth in human–computer interaction has led to a redefinition of requirements for effective user interfaces and that a key component of these requirements is the ability of systems to address affect. Learning a programming language is a complex cognitive process and it is argued that how people feel may play an important role on their cognitive processes as well (Goleman 1995). At the same time, many researchers acknowledge that affect has been overlooked by the computer community in general (Picard and Klein 2002). A remedy in the problem of effectively teaching children through educational applications may lie in rendering computer assisted e-learning systems more human-like and thus more affective. To this end, the incorporation of emotion recognition components as well as the incorporation of animated tutoring agents in the user interface of the educational application can be quite useful and profitable (Elliott et al. 1999). Indeed, the presence of animated, speaking agents has been considered beneficial for educational software (Johnson et al. 2000; Lester et al. 1997).

In the past, the authors of this chapter participated in the development of prototype systems that incorporate emotion recognition modules, based on artificial intelligence techniques and multi-attribute decision making approaches (Alepis et al. 2009, 2010). The resulting systems showed significant efficiency in recognizing and reproducing emotions.

After a thorough investigation in the related scientific literature we found that there is a shortage of educational systems that incorporate multi-modal emotion recognition, while we did not find any existing programming languages that incorporate emotion recognition and/or emotion generation modules. Perhaps the most relevant work is that of Kahn (1996), where an animated programming environment for children is described. The author of that paper has developed a programming interface called ToonTalk in which the source code is animated and the programming environment is a video game. The aim of that project was to give children the opportunity to build real programs in a manner that was easy to learn and fun to do. However, that approach did not incorporate any affective interaction capabilities.

In view of the above, this chapter presents a new programming language for children, which is highly interactive and intelligent since it provides affective interaction during programming. The programming language was named m-AFOL which is the acronym for "mobile-Affective Object Oriented Logo Language". In the implementation of the m-AFOL language, there is an added programming dimension that of object oriented programming (Pastor et al. 2001; Alepis and Virvou 2012). Through the language's object-oriented architecture, an m-AFOL program may thus be viewed as a collection of interacting objects, with their attributes and their methods, as opposed to the conventional Logo model, in which a program is seen as a list of tasks. Furthermore, the initial goal to create a programming language suitable for the needs and the limitations of children is further improved through the m-AFOL language, by a highly user-friendly user interface, designed for affective interaction between high-tech mobile phones (smartphones) and children. The mobile dimension in programming interfaces, the incorporation of emotion recognition capabilities, as well as the presence of speaking animated emotional tutoring agents all introduce novelties in the area of mobile learning.

10.2 General Architecture of the m-AFOL Programming Environment

In this section, we describe the overall functionality and emotion recognition features of m-AFOL. As a test bed for creating the mobile programming platform we have used the well-known Android operation system. The whole platform is written in Java programming language, while the databases that are used are MySQL databases. Both the programming language and the database are wide spread and free to use. Furthermore, Java programming language is perhaps the most well-known Object Oriented programming language and thus conforms to the requirements of the whole system's architecture. An outline of the general system's UML class diagram is illustrated in Fig. 10.1. Basic classes are illustrated along with selected attributes, operations, associations and generalizations. The basic classes are the m-AFOL system and the Student class. m-AFOL consists of four basic subsystems that control the whole platform's behavior with its interacting users. The central m-AFOL class is able to connect with a remote server who serves all mobile applications that are running m-AFOL. Some "heavy" processing operations such as algorithmic processes for emotion recognition take place on the server and their output is transmitted back to the client devices.

The architecture of m-AFOL consists of the main educational programming environment, a user monitoring component, emotion recognition inference mechanisms and a database. A part of the database is used to store data related to the programming language while another part is used to store and handle affective interaction related data. The programming environment's general process is illustrated in Fig. 10.2. The architecture of m-AFOL consists of the main educational application, a user monitoring component, emotion recognition inference mechanisms and a database. Part of the database is used to store educational data and data related to the tutoring agent. Another part of the database is used to store and handle emotion recognition related data. Finally, the database is also used to store user models and user personal profiles for each individual user that uses and interacts with the system. Figure 10.2 also illustrates the steps that are followed during each student-mobile interaction.

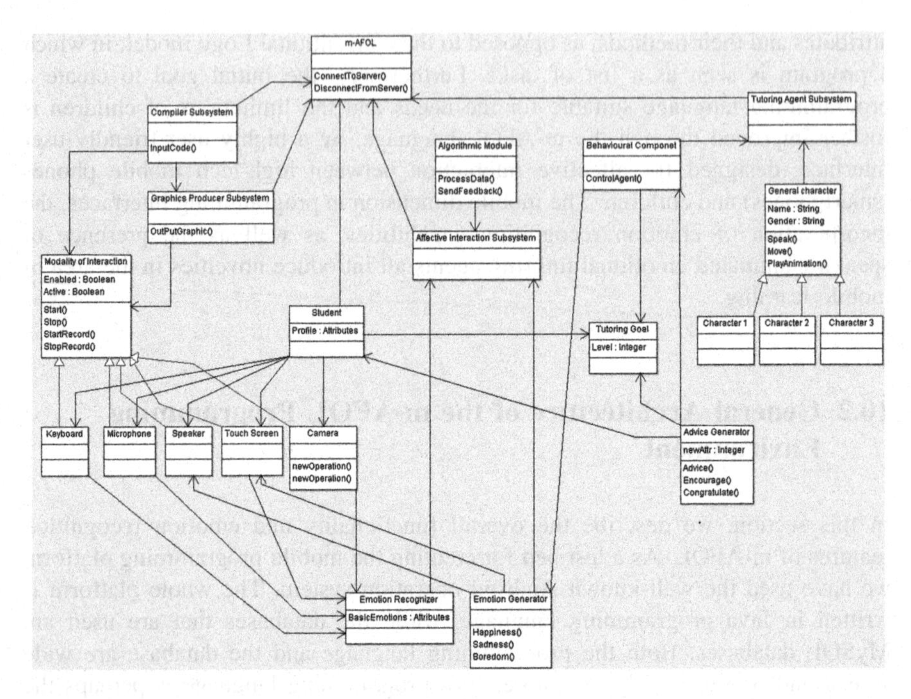

Fig. 10.1 UML Class diagram for the platform's main architecture

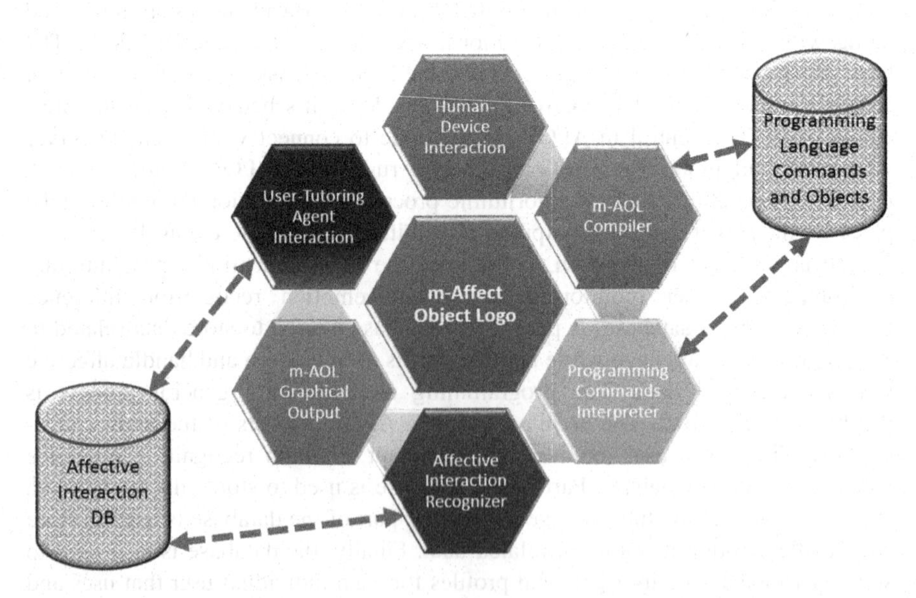

Fig. 10.2 General process in the m-AFOL platform

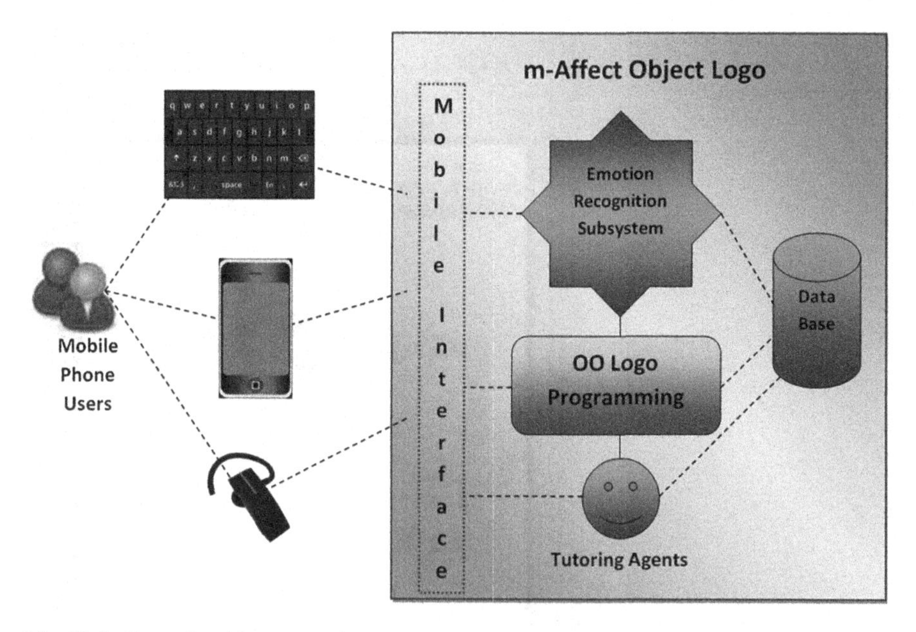

Fig. 10.3 General architecture and user interaction of m-AFOL

As we can see in Fig. 10.3, the students' interaction is multi-modal, which means that it can be accomplished orally through the microphone, through the mobile keyboard and through the mobile touch screen modality that is usually equipped with a front camera. The first three modalities of interaction, namely keyboard, microphone and touchscreen are used for inputting data to the platform. Mobile speakers and screens can be used as the interface where the platform outputs data to its users. Finally, a front camera can be used to capture videos and/or images of the operating user in order to be processed and used for facial emotion recognition. The educational process needs three basic subsystems to cooperate with each other, namely the affective interaction subsystem, the programming language's compiler subsystem and the subsystem that reasons for and handles the animated agent's behaviour. Both the programming language's interface and the tutoring agent are visible in the mobile platform, while the emotion recognition module is invisible as it serves as a reasoning mechanism responsible for the affective interaction between the mobile platform and its potential users.

Users of the m-AFOL mobile platform can choose whether they want to use the system's affective interaction capabilities. Furthermore, it rests on the users, whether they choose to enable or not the available modalities that can be used for emotion recognition. Figure 10.4 illustrates the mobile application's settings form, where users can choose to enable or disable the audio and the video emotion recognition module. In Fig. 10.4 both the audio and the video modality of interaction are enabled.

Fig. 10.4 Settings for emotion recognition modalities of interaction

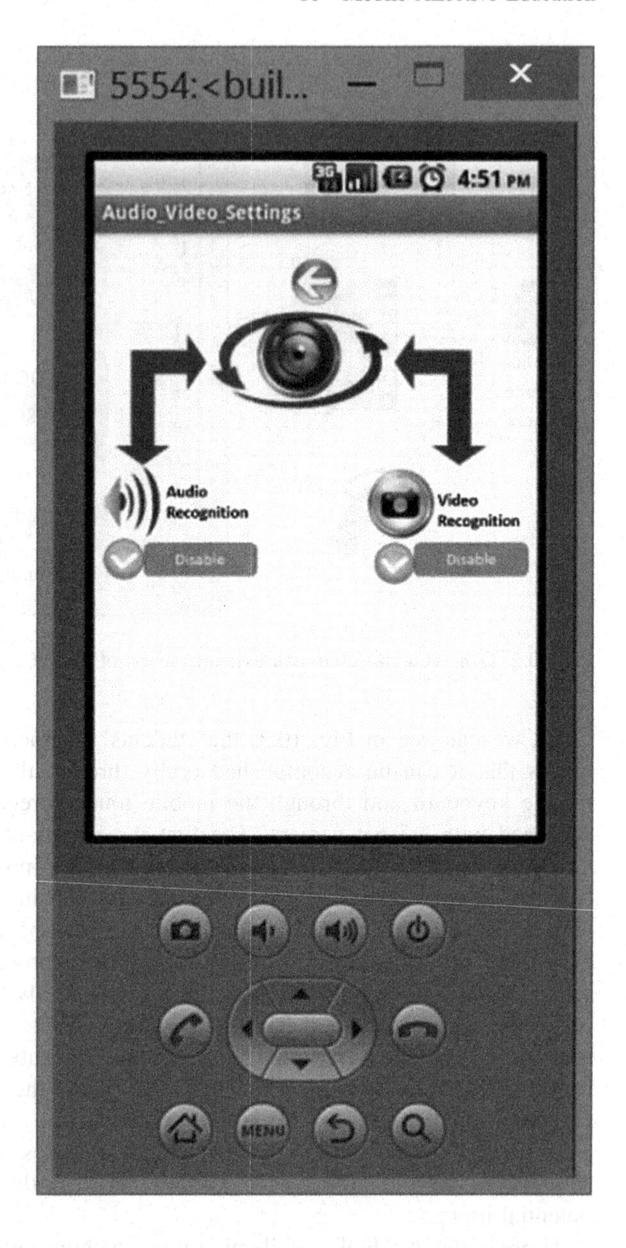

Figure 10.5 illustrates a snapshot where the video recognition module is enabled and corespondigly the mobile device's front camera is activated. As the readers may notice, the user's image is captured by the mobile application, while the user's face is detected and tracked through the whole interaction. In Fig. 10.5 the user's detected face is noted with a yellow square.

Fig. 10.5 Face detection and tracking during the interaction with m-AFOL

10.3 Overview of the m-AFOL Programming Learning System

While using the educational application from a smartphone, children as the platform's students have the oportunity to learn programming through programming courses. The information is presented in graphical form while an animated tutoring agent is present and may alternatively read the theory out loud using a speech engine of each user's choice. Students are prompted to write programming commands as first steps towards getting to know the educational platform. Advanced level students are given graphical illustrations as examples and are prompted to write complete programs in the m-AFOL language in order to produce drawings, shapes and particular objects. The mobile application is installed locally in each student's smartphone device. In the first phase of this project, the resulting application is targeted to the Android OS platform, while there are also future plans for integration to other mobile operation systems as well. Two examples of using the m-AFOL's programming interface are illustrated in Figs. 10.6, 10.7, and 10.8. The animated agent is present in these modes to make the interaction more human-like. It is worth noticing that in both examples the tutoring agent would also try to sympathise with the aims of each user trying to elicit human emotions.

Figure 6 illustrates a snapshot from user interaction where a student is typing programming commands in order to produce a complicated drawing which is illustrated in Fig. 10.7. In Fig. 10.8 we can see a tutoring character that congratulates a user for successfully producing a quite complicated drawing by using the programming language's commands and object oriented features. As a reader may notice in Fig. 10.6, the programming commands are given in a more "procedural" way. It is true, that while learning basic programming principles, it is quite easier to teach "simple" procedural thinking by writing simple commands. However, in OO programming, code becomes more "elegant" and "powerful". After completing the first level courses, students using m-AFOL are taught basic OO programming principles and are then encouraged to use them in their programs. As a result, students can produce more complicated drawings with fewer lines of code, or even the same drawings (made with the procedural way), but with substantially less effort. Of course, there is a very important precondition that they should have firstly mastered the basic OO programming principles and concepts in order to use everything this programming paradigm has to offer.

To this end, Figs. 10.9 and 10.10 illustrate programming using m-AFOL. More specifically, writing simple procedural commands is illustrated in Fig. 10.9, while OO programming by calling objects and their methods is illustrated in Fig. 10.10. The OO interface is expected to be more powerful and user friendly for young students. Their graphic creations can be saved not only as pictures but also as autonomous programming code sets. In this way, their code becomes reusable and extendable for their future needs (two more benefits in OO programming).

While the students interact with the mobile affective system, their entire oral, keyboard, on-screen and optionally visual actions are recorded in order to be interpreted in emotional interaction terms. An in-depth description of the affective module is beyond the scopes of this chapter. However, the overall functionality of these modules which lead to multi-modal recognition of emotions through multimodal data, can be found in Tsihrintzis et al. (2008) and in Alepis and Virvou (2006). The results from these studies were quite promising and encouraged the authors to test this approach on a more demanding domain that belongs to the area of teaching programming to young children.

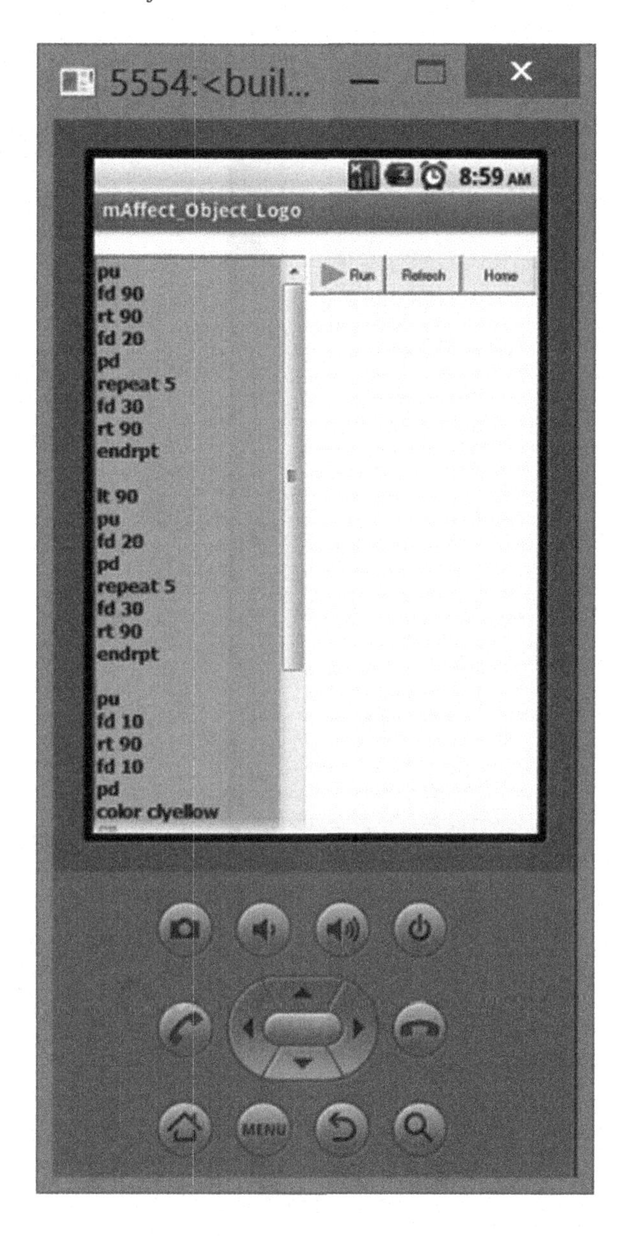

Fig. 10.6 A student is typing programming commands in his/her smartphone

10.4 m-AFOL Language Commands and Object Oriented Structure

In this section we give an overview of the m-AFOL system's supported programming commands, as well as the functionality that derives from the language's Object Oriented architecture. In the implementation of the m-AFOL system we

Fig. 10.7 A students executes an m-AFOL program to produce a specific drawing

have used the past Logo's well-known feature character which is a turtle. The turtle is an on-screen cursor, which can be given movement and drawing instructions, and is used to programmatically produce line graphics and colored shapes. Programming code snippets that produce drawing objects can be saved and stored as Objects of the m-AFOL system. Correspondingly, stored objects can be reloaded and used within the programming language as existing pieces of programming code.

Fig. 10.8 The animated agent is congratulating a student

Tables 10.1 and 10.2 show indicative elements of the m-AFOL's objects' structure and the programming language's commands respectively.

Table 10.1 illustrates a subset of the attributes and the methods that can be used/called within the Object Oriented structure of the m-AFOL language. Each existing object can be called by its name in the system's command-line interface. Correspondingly, each specified object has a number of attributes and operations

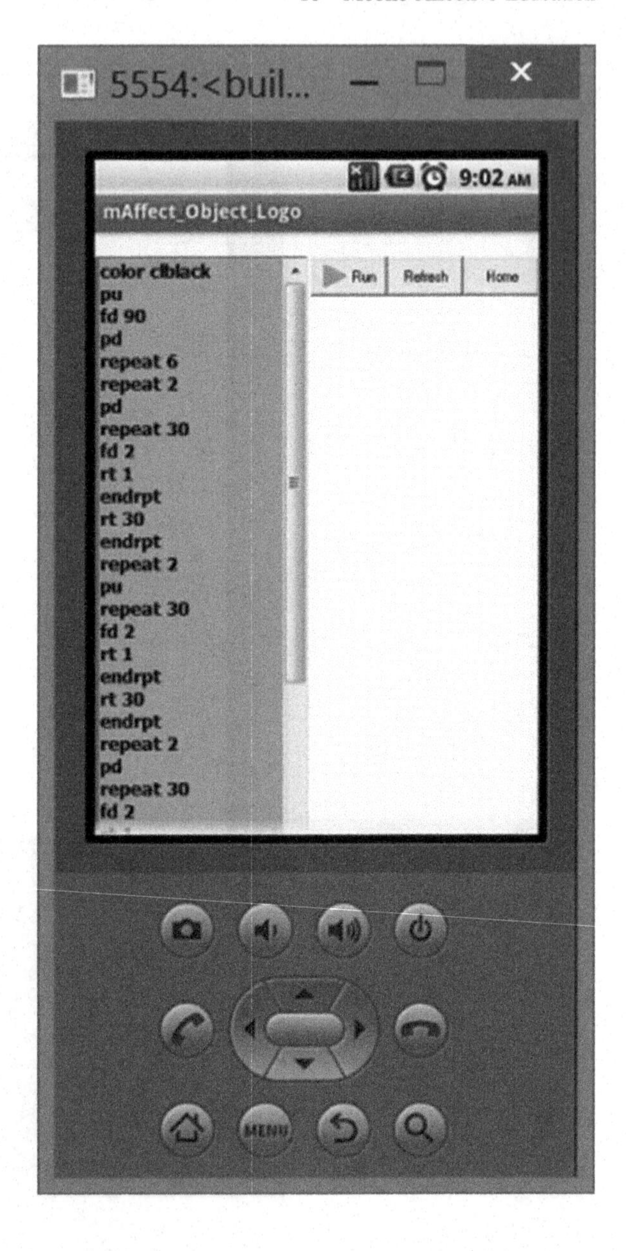

Fig. 10.9 Procedural programming in the m-AFOL user interface

that can take values or can be called on run time by the users. Both the attributes and the operations constitute each objects characteristics and functionalities and are members of a general template class. In accordance with the implementation of the m-AFOL language, if an object consists of other objects (for example a drawn house may contain drawn windows and doors), then these sub-objects also inherit the super-object's characteristics. However, all inherited characteristics are the

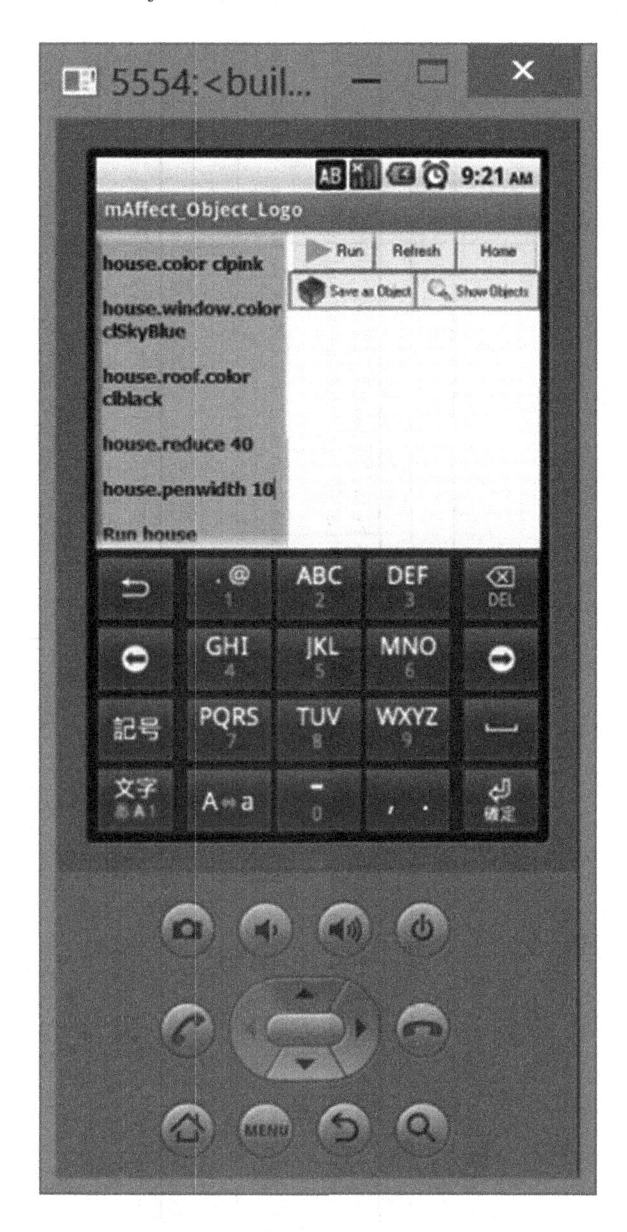

Fig. 10.10 OO programming in the m-AFOL user interface

default values for each sub-object and these values can be changed in each sub-object's own implementation.

At this point we should state that each object's attributes provide an alternative way of assigning values while drawing in the m-AFOL's interface. For example by writing "NewHouse.color clgreen" we instruct the m-AFOL's turtle to use the green color when drawing the "NewHouse" object. In this case the precondition is

Table 10.1 m-AFOL language object attributes and methods

Object attributes	
Attribute	**Description**
Color	This attribute assigns a color value to the specified Object. The color value can be used by the "drawing" and the "filling" commands
Penwidth	This attribute assigns a penwidth (width of pen) value to the specified Object. The width of the pen value can be used by the "drawing" commands
Pendown	Boolean attribute. Indicates whether the platform's turtle is able to draw while moving (value set to true) or not (value set to false). Default value is True

Object operation	
Operation	**Description**
Reduce	This operation is used to reduce the size of the calling specified Object by a user specified percentage. Affects by the specified percentage all forward and backward turtle drawing movements
Expand	This operation is used to expand the size of a specified Object by a user specified percentage. Affects all forward and backward turtle drawing movements
Flip horizontal	This operation is used to flip horizontally a specified Object. Affects all right and left turning commands
Flip vertical	This operation is used to flip vertically a specified Object. Affects all right and left turning commands
Rotate	This operation is used to draw a specified Object rotated clockwise by a user specified angle. Affects the turtle's initial angle
Mirror	This operation is used in order to produce a drawing that is the mirroring object of the calling object
Save	This operation is used to save a specified Object with its name in the system's database
Load	This operation is used to load a specific Object by its name. This object must be pre-stored in the system's database

Table 10.2 m-AFOL language basic commands

Affective object logo commands	Description of command	Example
Fd *value*	This command is used to move the turtle forward by a specified by the "value" variable range (move forward)	Fd 100
Bd *value*	This command is used to move the turtle backward by a specified by the "value" variable range (move backward)	Bd 80
Rt *value*	This command is used to rotate the turtle clockwise by a specified by the "value" variable range (turn right)	Rt 45
Lt *value*	This command is used to rotate the turtle counterclockwise by a specified by the "value" variable range (turn left)	Lt 180
Fill	This procedure is used to fill a closed shape with the turtle's current color. If the shape is not closed this command will paint the whole screens drawing area with the specified color	Fill
Color *colorvariable*	This command is used to change the turtle's current drawing color to the specified by the "colorvariable" variable color	Color clBlue
PenWidth *value*	This command is used to change the turtle's current drawing pen width to the specified by the "value" variable value	PenWidth 3
Pu	This command is used to hold the turtle's pen up, so that the turtle can move without leaving any trace (pen up)	Pu
Pd	This command is used to restore the turtle's pen down (after a pu command), so that the turtle can move leaving its trace (pen down)	Pd
Refresh	This command is used to refresh/clear the drawing area	Refresh
Home	This command is used to move the turtle to a default position in the center of the drawing area	Home
New *Object*	This command is used when creating a new Object by inheritance. The new Object inherits its functionality by a pre-stored mother Object to the system's database Object	NewHouse SmallHouse
Delete *Object*	This command is used in order to delete a pre-stored or on run time created object	Delete NewHouse
Run *Object*	This command is used to run all the commands that constitute the programming code of a specified pre-loaded Object	Run NewHouse
Repeat *value*	This command is used to specify the beginning of a loop. Each loop contains commands and is repeated by a number of repetitions specified by the "value" variable	Repeat 4 Fd 50 Rt 90
Endrpt	This command is used to indicate the end of a "repeat" loop	Endrpt

that this object is already created and stored in the languages database and also that this object is "loaded" first by calling its name. The specific color operation can be also alternatively achieved by assigning the red color as the value of the turtle's current drawing pen before drawing the specified object. However, each object's methods have a more complicated structure and it would be quite difficult to provide an alternative way for their implementation rather than their calls through the existing objects. Furthermore, calling objects with their methods and their attributes provides more readable and understandable programming code, something that is appreciated both by novice and expert programmers. As another example, if we wanted to reduce the size of a specified object, the "reduce call" would easily do that ("newobject.reduce 75", which means that the new object will have 75 % of the size of the initial object), while a user would have to change a large number of commands in the object's implementation in order to have the same effect. As a result, the OO structure within the m-AFOL's commands not only provides a better, logical structure for the existing programming language, but also offers more functionality to the users.

Table 10.2 shows a listing of the m-AFOL programming language interface commands, as well as an example of programming code for each command. In the m-AFOL's GUI the turtle moves with commands that are relative to its own position, for example Rt 45 means rotate right by 45 degrees. Students who use the m-AFOL system would easily understand (and predict and reason about) the turtle's motion by imagining what they would do if they were the turtle. Some commands produce a drawing effect while the turtle moves on the system's screen, while other commands are used to handle objects and to refresh the drawing area. Furthermore, some commands affect the turtle's available pen that is used for drawing a trace when the turtle moves. Finally, the "repeat" command is used to include a sequence of commands that is executed repeatedly for a specified number of times.

10.5 Conclusions

In this chapter, we presented a new programming language that has been implemented for the needs of teaching basic programming skills and principles to young children. The resulting sophisticated programming environment is called m-AFOL. The resulting programming environment reflects the authors' attempt to preserve well-known past approaches towards the education of children in the area of programming, while at the same time enrich the language's architecture with modern architectures and approaches and also provides a highly attractive interface for young children. The "older" Logo programming language of the past has been used as a foretype for the prototype system which has been extended by the incorporation of the wide spread paradigm of object oriented programming. The programming language's environment is also very user friendly since it includes affective interaction components, namely multi-modal emotion recognition and emotion elicitation through interactive tutoring agents that participate in the whole educational process. Mobile interaction with the entire programming interface is also achieved for high-end mobile devices known as smartphones and initially for the Android OS.

It is part of our future plans to extend the m-AFOL system to more operational systems. Moreover, a future evaluation study is expected to reveal specific affective characteristics of the mobile-learning environment that may influence the children in becoming more effective students and more satisfied users.

References

Alepis E (2011) AFOL: towards a new intelligent interactive programming language for children. In: Smart innovation, systems and technologies (SIST), vol 11. Springer, pp 199–208

Alepis E, Stathopoulou I-O, Virvou M, Tsihrintzis GA, Kabassi K (2010) Audio-lingual and visual-facial emotion recognition: towards a bi-modal interaction system. In: Proceedings of the international conference on tools with artificial intelligence (ICTAI), vol 2. Article number 5670096, pp 274–281

Alepis E, Virvou M (2012) Multimodal object oriented user interfaces in mobile affective interaction. Multimedia Tools Appl 59(1):41–63

Alepis E, Virvou M, Kabassi K (2009) Recognition and generation of emotions in affective e-learning. In: Proceedings of the 4th international conference on software and data technologies (ICSOFT 2009), vol 2. pp 273–280

Alepis E, Virvou L (2006) Emotional intelligence: constructing user stereotypes for affective bimodal interaction. In: Lecture notes in computer science: knowledge-based intelligent information and engineering systems, LNAI I, vol 4251. Springer, Berlin, pp 435–442

Elliott C, Rickel J, Lester J (1999) Lifelike pedagogical agents and affective computing: an exploratory synthesis. In: Wooldridge MJ, Veloso M (eds) Artificial intelligence today. LNCS 1600. Springer, Berlin, pp 195–212

Feurzeig W (2010) Towards a culture of creativity: a personal perspective on logo's early years and ongoing potential. Int J Comput Math Learn, 1–9 (Article in press)

Frazier F (1967) The logo system: preliminary manual. BBN technical report. BBN Technologies, Cambridge

Goleman D (1995) Emotional intelligence. Bantam Books Inc, New York

Hudlicka E (2003) To feel or not to feel: the role of affect in human-computer interaction. Int J Human–Comput Stud Elsevier Sci 59:1–32 (London)

Johnson WL, Rickel J, Lester J (2000) Animated pedagogical agents: face-to-face interaction in interactive learning environments. Int J Artif Intell Edu 11:47–78

Kahn K (1996) ToonTalk—an animated programming environment for children. J Vis Lang Comput 7(2):197–217

Lester J, Converse S, Kahler S, Barlow S, Stone B, Bhogal R (1997). The persona effect: affective impact of animated pedagogical agents. In: Pemberton S (ed) Human factors in computing systems. Proceedings of the CHI' 97 conference. ACM Press, pp 359–366

Pastor O, Gómez J, Insfrán E, Pelechano V (2001) The OO-Method approach for information systems modeling: from object-oriented conceptual modeling to automated programming. Info Syst 26(7):507–534

Picard RW, Klein J (2002) Computers that recognise and respond to user emotion: theoretical and practical implications. Interact Comput 14(2):141–169

Tsihrintzis G, Virvou M, Stathopoulou IO, Alepis E (2008) On improving visual-facial emotion recognition with audio-lingual and keyboard stroke pattern information. In: Web intelligence and intelligent agent technology (WI-IAT '08), vol 1. pp 810–816

Chapter 11
Conclusions

Abstract In the last chapter of this book the authors present their conclusions, derived from their quite recent research studies over mobile technology and mobile software. More specifically, their efforts were targeted to the domains of mobile learning, user adaptivity and multimodal mobile interfaces. Their proposed architectures and methodologies have concluded in building actual software systems and platforms, which were also evaluated both by computer specialists and also by real end mobile users. This book's last discussion reveals a fulfillment in the authors' attempts in the related scientific fields, as well as their suggestions and challenges for future pursuits in mobile learning software.

11.1 Conclusions

Mobile phones have become very popular among people and thus they are imposing a new culture where smartphones are part of peoples' everyday life. As a result, there is strong evidence that their use in computer supported education as a new tutoring and communication medium can be very useful both for the tutors and also for the learners. By enabling learners' access to educational content anywhere and anytime, mobile learning has both the potential to provide online learners with new opportunities and to reach less privileged categories of learners that lack access to traditional e-learning services (Moldovan et al. 2013). Consequently, schools will not only need to evaluate their school curriculums but also recognize the power in the digital devices to engage, enable and empower mobile learners (Keengwe et al. 2012). However, in the case of education, many design issues have to be taken seriously into account so that the resulting mobile software can be educationally beneficial to students and be included in the educational process. Among these important design features is the high degree of adaptivity and personalization that has to be achieved in these systems. These features should include student modeling facilities that are mainly used by modern ITSs.

E. Alepis and M. Virvou, *Object-Oriented User Interfaces for Personalized Mobile Learning*, Intelligent Systems Reference Library 64,
DOI: 10.1007/978-3-642-53851-3_11, © Springer-Verlag Berlin Heidelberg 2014

Through this book the authors aimed at presenting and describing solutions that derive from the Object Oriented paradigm for problems lying in the area of mobile learning. These problems, at the current state of technological means, cannot be treated as simple, as per evidence from a significant number of researchers and computer scientist from all over the world. More specifically, the authors have constructed a mobile authoring tool in Chap. 4, which they have evaluated and compared desktop software in Chap. 8. A mobile platform supporting students with special needs was discussed in Chap. 5. Two more platforms where presented in Chaps. 7 and 9 respectively. Both platforms targeted in addressing affects in mobile educational applications. For this purpose, multi modal interaction data were handled.

In this book, we have presented multimodal mobile user interfaces that support affective interaction in mobile educational applications. The incorporation of emotion recognition capabilities in mobile educational system is quite difficult and demanding and such efforts have not been found yet in related scientific literature. In order to accommodate this incorporation, we have used the OO model for the architecture of a remote emotion detection server that is responsible for the affective interaction, while mobile devices are served as clients. All connected, to the remote server, mobile devices exchange their data wirelessly and whenever it is required, data from the emotion recognition process are transmitted back to the mobile devices.

All mobile systems described in this book use the OO approach to use and handle evidence during human-mobile device interaction. The interaction's evidence combined with data from user models are then classified into well-structured objects with their own properties and methods. The OO model provides flexibility also for the mobile affective systems, since new users, new mobile devices and new modalities can be easily added and successfully incorporated. The OO architecture provides a structure for the available affective interaction data that makes them usable and exploitable for the emotion detection algorithms that try to classify them. This structure also provides easiness in the resulting systems' maintenance, great extensibility, better communication through different modalities, good cooperation with different object oriented programming languages and with different algorithms and lastly, an easiness in code debugging, as well as code reusability.

Challenges in future mobile educational software include platform and device independence, user friendlier mobile interfaces, low energy and less processing power consuming software applications and higher mobility in terms of hardware and mobile networks. As the economic crisis in all over the world grows, mobile software solutions and services of high technology may open new ways of deployment and economic growth. According to (Alizadeh and Hassan 2013), common problems encountered by Mobile Cloud Computing are privacy, personal data management, identity authentication, and potential attacks. Another review discussing such challenges can be found in (Pierre 2001), where it is stated that major challenges in mobile computing include mobility, disconnection and scale, new information medium and new resource limitations. Finally, in (Casany et al. 2013) we can find an interesting review about challenges in mobile learning. The authors suggest that two key challenges of mobile learning in developing countries is to incorporate the latest computing technologies in mobile devices and to guaranty that the quality and cost of Internet mobile access is independent from the kind of service that is used and intended for.

The results of the conducted evaluation studies that are presented in this book have been very positive and promising and have provided strong evidence that even in more demanding computerized situations as in mobile software, high level human-mobile device interaction can be accomplishable. It is the authors' strong belief that mobile education will continue serving peoples' need to learn in even more efficient and effective ways in the years to follow.

References

Alizadeh M, Hassan WH (2013) Challenges and opportunities of mobile cloud computing. In: 9th International wireless communications and mobile computing conference (IWCMC 2013), Article number 6583636, pp 660–666

Casany MJ, Alier M, Mayol E, Conde MA, García-Peñalvo FJ (2013) Mobile learning as an asset for development: challenges and oportunities. Commun Comput Inf Sci 278:244–250

Keengwe J, Schnellert G, Jonas D (2012) Mobile phones in education: challenges and opportunities for learning. Educ Inf Technol pp 1–10

Moldovan A-N, Weibelzahl S, Muntean CH (2013) Energy-aware mobile learning: opportunities and challenges. IEEE Commun Surv Tutorials (in press) (ISSN 1553877X) DOI: 10.1109/SURV.2013.071913.00194

Pierre S (2001) Mobile computing and ubiquitous networking: concepts, technologies and challenges. Telematics Inform 18(2–3):109–131

Printed in the United States
By Bookmasters